Climbing the Vines
in Burgundy

Climbing the Vines in Burgundy

How an American Came to Own a Legendary Vineyard in France

Alex Gambal

ROWMAN & LITTLEFIELD
Lanham • Boulder • New York • London

Published by Rowman & Littlefield
An imprint of The Rowman & Littlefield Publishing Group, Inc.
4501 Forbes Boulevard, Suite 200, Lanham, Maryland 20706
www.rowman.com

86-90 Paul Street, London EC2A 4NE, United Kingdom

Distributed by NATIONAL BOOK NETWORK

British Library Cataloguing in Publication Information Available

Library of Congress Cataloguing in Publication Information Available

LCCN: 2023941330

♾™ The paper used in this publication meets the minimum requirements of American National Standard for Information Sciences—Permanence of Paper for Printed Library Materials, ANSI/NISO Z39.48-1992.

To Diana, mon petit chou

Contents

PART 1

L'Enracinement

Taking Root

Vines in Burgundy are planted in late winter to benefit from the wet spring. The scion, or fruit, is grafted onto rootstock that originated from wild US vines. The soil has been protected by Burgundy's winter cloak of fog. After the solstice, as the daylight returns and the air grows warmer, we complete the last of the pruning and burn the cuttings. Winter crops, planted last fall, come out of their dormancy and rise up in the fields. Colza, from which we get canola oil, is planted, will later flower, creating fields of yellow. Cold rains can hinder vineyard work but, this rain is a reservoir for the soil and its storage allows the vine to flourish until harvest. As winter retreats, wildflowers appear and buds on the cherry trees begin to break through.

Chapter 1

La Haute Plongée (The High Dive)

"You did what?" You would think that after all this time, the shock would wear off. But this is still the reaction I get when I tell people that I left the security of a successful career in my family business in the States to move to Burgundy and become a vigneron (winemaker). In fact, everyone I meet since I jumped off this high dive looks at me quizzically, with a mysterious smile indicating incredulity or perhaps terror.

In 1987, I was 30 years old. My father's business partner gave me a mixed case of wine from the local wine retailer Mayflower Wines and Spirits for Christmas. Each month the owner, Sidney Moore, a woman then in her 50s who carried a bohemian Joan Baez vibe, would offer assorted cases of wines that she had directly sourced, accompanying them with copious notes. This was when retailers knew how to sell. There were no tweets, no Instagram, no email blasts. Back then, faxes were just appearing. Robert Parker, the wine critic, was creating a career for himself just miles north of me in Baltimore. Newsletters were snail mailed and written in a lively style that made the recipient eager to devour them line by line. Much like an old style "salon" where you gather to discuss an interesting subject, this was marketing in its pure form: written to titillate the customer.

Fascinated by Moore's newsletter, I saw she was offering wine tasting classes on Fridays after work. I signed up. I do not remember all that I learned, but because of these classes I now understand the questions I hear from beginners. When you begin to taste and drink wine in earnest, you are learning a new language. This language, without sounds, is about taste, smell and texture. It is overwhelming at first and takes work. Pleasurable work to be sure, but work nonetheless.

The regions, the grapes, the producers; what am I smelling, what am I supposed to smell, what is this about color and what's with the legs? (Forget the legs, absolutely unimportant unless you want to sound pretentious.) As with most new languages, it is laden with secrets. A class, yes education, helps you pry open a heavy old hatch that must be lifted and secured. You then can peer

3

inside and take measure of what you find, like entering a wine cellar for the first time. The customs and mysteries associated with the language of wine make it more like a secret society than a drink. It is not really, but most of the wine trade would like to maintain this façade in order to justify their power as arbiters.

Sidney's store, at New Hampshire Avenue and M Street, was a welcoming storefront a couple of blocks from my office in Washington, DC. It was not slick. No cedar racks. No frosted glass. But it was not a dusty old dungeon. It was, to borrow a title popularized by Ernest Hemingway, "A Clean, Well-Lighted Place." The shelves were well organized with cases stacked on the floors or in the loft above the sales floor. Sidney and her son Harry paired their desks in the back on a slightly raised platform. At lunchtime, I would grab a sandwich next door and head to the back of the wine store to chat with them about what she had cooked over the weekend, new restaurants and tastings on the horizon. Only after I ate my lunch would I get the gumption to ask, "do you have anything open?" Often it was a bottle that had been opened for several days and she would say "taste at your own risk."

Retailers in Washington, DC, can import directly, making the region a wine freak's Toys "R" Us. Choice and transparency have become the operative words throughout the US over the last thirty years with the growth of the internet and power of large chain retailers such as Total Wine, Costco, and Whole Foods. Washington, however, was a pioneer and still is with four-to-five wonderful independent stores that are true wine specialists. This is a legacy of the District's peculiar liquor laws, no doubt set up to benefit Congress members who often came from dry districts to the great benefit of the entire Washington metropolitan area. Our nation's history is odd when it comes to wine and spirits. It was not until 1968 that you could get a drink in Virginia. Even today, liquor still has to be bought at Virginia state stores and the affluent Maryland suburb of Montgomery County controls the sale of wine and liquor through county stores. Thus, Washington, DC, became an island of choices for wine lovers.

Another main attraction to Moore's store was the specialty foods Sidney imported from Italy via New Jersey. Each Thursday or Friday she would have a shipment of crusty bread and flat bread known as panelles, small, medium, and large shipped down from an Italian baker in Hoboken. My taste buds water when I remember the smell of the loaves piled high at the entrance of the store coated with fresh oregano and cheese. Her mother exported homemade tomato sauce from Tuscany, green olives, olive oil from Badia a Contibuono, artichokes and stuffed peppers preserved in oil. Back then what she was doing was revolutionary in Washington, DC. Of course, there was Zabars in New York and a handful of old-style local Italian, Greek, and

Kosher ethnic grocery stores in South Philly and other Northeast cities, but Whole Foods was a long way off.

I was starving for a new experience, getting itchy from the routine of managing parking lots, garages, and real estate development, my father's self-built family business. My wife, Nancy, a physical therapist, was a passionate artist at heart. She began to take night classes in graphic design. We had two young children, Tyler, 5, and Alexa, 2, so it was not as if Nancy and I were going out to clubs and discos. Ours was a social life of what are we having for dinner, what are we drinking, who is putting the kids to bed, and who is reading to them? Many of our friends were in the same stage of life, so what did we do? Made dinner and drank wine. We popped corks at afternoon BBQs and told the multitude of children to "go play." The moms and dads daydreamed while drinking wine. In the evenings if a babysitter was available, we would rotate our tastings from house to house arranging soirées around a theme or a wine region. For most of the group, wine was a distraction. But for me, it became an obsession. I wanted to know more; I was intrigued about the lives of the winemakers, their stories, the geology, and I could remember what I tasted.

Along with the gourmet dry goods at Sidney's there was, of course, racks upon racks of wine. Her father brought in gems from Italy that had been largely unknown. They were the first to import small producers of Barolo and Barbaresco from the Piedmont and Chianti from Tuscany going back to the 1960s and the 1950s. I remember the '57s and '58s being spectacular. These were amazing wines, a fabulous story themselves. It not only made me want to know the wines but meet the winemakers. Who were these artisans?

These bottles were time capsules that appeared at the edge of an era when a new generation of winemakers arrived and cleaned up what was arguably an industry full of some pretty filthy wine. In all honesty, the same argument can be made for the new generations that were transforming California, France, Australia, and New Zealand. A wave of baby boomers was either taking over from their parents, or had come to the profession independently, with new ideas, often with degrees in oenology and viticulture. Many had worked in other regions and were swept up in a winemaking renaissance that started in the mid- to late 1980s and that continues today.

I hate to sound like an old geezer, but when I began to taste and buy wine in the late '80s, there were, simply put, some poorly made and undrinkable wine on the shelves. Today it is very difficult to find a poorly made bottle, but what is difficult is to find a wine made passionately, with blood, sweat, tears, and love.

I did not have a legacy or even a natural inclination for wine, but I did grow up with wine on the table. My adolescent mornings in Alexandria, Virginia, began with a fairly typical fare of slightly burnt toast, bacon, Cap'n Crunch and the ubiquitous smell of my mother's cigarette. A far too skimpy lunch at a school for boys—I could have eaten two to three more helpings—followed by sports. I came home by 5:30 pm absolutely starving.

My mother was not a great cook. But she knew how to put good food and plenty of it on the table, especially since our home tended to be the safe house for a mélange of boys and girls once we got our drivers licenses. Pots of meaty spaghetti sauce, chicken soup, and thick stew, Mom's version of Shake 'N Bake chicken, fully dead pot roast, and the famous Gambal "bad hens," Perdue Cornish game hens that my mother bought fresh, froze, and then cooked to within an inch of their lives. The list goes on. It all tasted so good and got us to sit at the table so our mother could keep an eye on us and find out what was going on in our lives. (Years later in France, I discovered that the so-called Cornish game hen is nothing more than a young chicken, less than twenty-eight days old, female or male, and in France called a poussin or coquelette. It is one of my favorite dishes.)

By the time my father got home from work, we were ready to eat our arms off; we were brain dead, tired, fussy, and sick of eating crackers and being told you will ruin your appetite. First, Dad needed to fix Mom's Canadian Club and soda. Dad popped open his beer and brought the wine to the table. When I was around 11 or 12 years old, I asked if I could open the wine. Our opener was flimsy with the two handles you pull down to remove the cork. I loved the small pop, even with the risk of tearing my hand open because of the two lousy handles. We had three house regulars: Ruffino Chianti Classico and Mouton Cadet red and white. Occasionally at my uncle's for Thanksgiving in New Jersey, we might have a Bordeaux, but it was often Lancer's. That was my wine education until college when we started to drink the gallon jugs of Paisano and let's not even begin to talk about the next morning.

From these modest beginnings, the chance encounter with a mixed case at Christmas and through Sidney's salon, I met and was welcomed by a group of her customers into the inner sanctum of wine. I dislike the word collector because it implies a certain hoarding and one-upmanship but this "Fab Four" were true connoisseurs, from the French word conaissance or knowledge. They knew and understood the history, personality, and most importantly the aesthetic of wine, and they loved sharing this with me. My four mentors haunted the auction world when it was necessary to have real knowledge, knowledge that was gained by sharing as there were no wine chat rooms, no blogs, etc. It was dinner, a handful of great bottles and equally passionate and thoughtful discussion. Sure, there was excess, that was half the fun, but it was their ability to share that allowed me to begin to see inside the wine.

By 1991, the wine bug bit me hard. I was having an early mid-life crisis. I was good enough at my work in the family real estate and parking business, but I began to realize I wanted to create something of my own. I yearned to build something from the ground up, like my father, Serge. Dad was a tough guy who became a classic post-World War II entrepreneur. The grandson of Russian and Polish immigrants from the 1870s through the 1880s, he was born in Old Forge, Pennsylvania, next to Scranton, and came from a tough stock—coal miners most of them, plus Russian Orthodox priests. His mother, Alice, my Babushka was the toughest of them all. She ran the corner bodega, "The Gambal Market," selling penny candy and one cent cigarettes during the depression.

Dad was thrust into the Army in 1942, did two years training in Hattiesburg, Mississippi before being deployed (think Neil Simon and *Biloxi Blues*) and, as most of the fathers who were in combat, rarely talked about it (he was with the 69th Infantry Division that captured Leipzig, liberated a concentration camp, and met the Russians at Torgau on the Elbe). He did tell me once in all seriousness that he was glad to get out of Mississippi and go to Europe; getting shot at could not be worse than two years of mosquitos, canvas tents, and heavy khakis in the Mississippi summer.

After the war, the GI Bill allowed him to go to George Washington University and GW law school. With a year left in law school, he and his fraternity buddy, who had split duties running a parking lot in Foggy Bottom near the State Department and Watergate building, said to their parents they were quitting law school and going into the parking business. I do not envy the fly on the wall who heard my dad's parents ask him "You are doing what? You have a career track, you are the first person in our family to go to college, you are going to be a lawyer."

Starting with a single parking lot, they built the business with handshake deals, honest accounting, service, and always reinvesting in the business to become the largest parking company in the Washington, DC, area. They often said that the best decision they ever made was to leave law school, because they were more successful than they would have ever been as lawyers. Quite simply, neither my father nor his partner had the personalities to work for others.

My experience in the family business started early when in the summer of 1970, I could work a half day. I would drive in with my father to audit car tickets and cash receipts and take the bus home. Over the next several summers I cleaned parking lots and at 16, with driver's license in hand, I parked cars for minimum wage and tips in the delightful Washington, DC, summers. The tips were great. By the end of the week I would have a big bag full of quarters totaling $100. The city was my backyard, and the company was my family. I was comfortable and happy in the environment even though the expectation was that as Serge's eldest son, I would one day join the business.

Fast forward twenty years and that comfort level felt like a long, boring road ahead. Reality interceded in the form of a major real estate recession known as the Savings & Loan Crisis. I learned in my early 30s to do "workouts," which is a nice way of saying trying not to get stiffed by your partners who cannot cover their debts.

One gray, dog-dripping-wet November day in Washington, DC, I slogged from meeting to meeting with the family lawyer, the banks, and our partners. It was a hard lesson to realize there's not many people you can count on in a crisis. I saw the nakedness of risk and obligations: there were those who manned up and those who ran away.

At the end of that sobering day, I toiled in the door of my home in Alexandria, Virginia, and even the happy faces of my wife and children couldn't brighten the feeling I had. I was at end of a sealed-off tunnel. After the kids were in bed, I said, "There has to be something else we can do. We do not have to be here. I used to like my work. I am not indentured to my family, this is our life. I have lived in Alexandria all my life. There has to be somewhere else we can go, work, and have fun with our children."

"You know I am in with you on whatever you want you to do," Nancy replied. "This is our life. We do not have to stay here. We can choose wherever we want to be. We both love Italian wine and the food. What if I start Italian lessons and we explore the possibilities of leaving here and moving abroad?"

The question with a capital L was how to Leave? This was one of those horrible, circular Edgar Allan Poe–like moments where you are trapped in your own faulty logic, thinking that there is a moment that you will be saved or that you will discover a secret escape passage. We wait for the perfect moment to arrive, but it never does because we have not prepared.

I turned to a friend who was a year or two older than me and exuded a serious confidence, worldliness, and gravitas that I admired. We met in the early 1980s when he was a rising star with the global architectural firm Skidmore, Owings, and Merrill and was the architect of an office building that we built in Washington. I admired his work on our project and others that helped transform the old downtown from one of strip clubs to a lively neighborhood with restaurants, shops, and offices. We would often talk at lunch or in my office about what we wanted to do when we grew up. Our rhetorical "what ifs" led to "how to" conversations: How do you leave? How do you start new? How do you say you are leaving? How do you reinvent yourself when you have the fear of a 9-year-old at the ladder of the high dive? Do we climb the ladder? Do we walk to the edge of the board? We are at a fulcrum that can go in either direction. We retreat, but why? Bloody heck scared is the answer, but why? Why are we scared and how do we step off?

In the spring of 1991, we ate sandwiches in my office and looked at the land-use maps of Washington on the walls.

"In 1988, Frank Gehry told me that one day at a weekly meeting, one of his partners arrived and tacked to the bulletin board four plane tickets to Europe one year hence. The partner, at the risk of shock, used this as an unconventional way of saying 'we are leaving.' The tickets were for him and his family, and he explained that he knew that if he never performed the physical act of buying the tickets he would never leave. The act of buying the tickets was a commitment for him and his family that put in motion a chain of steps that would create a path to the airport."

In the summer of 1991, he left SOM to form his own firm in New York City specializing in smaller, aesthetically challenging projects, inspiring me to buy a plane ticket: but to where?

Chapter 2

Une Salle Pleine de Myrtilles
(A Room Full of Blueberries)

Sidney gave us the best piece of advice possible: "Don't think that you are going abroad to write the next great ex-patriot book. Don't expect it to be like *A Year in Provence*. If you want a great experience, you need to get a job, live in the country, learn the language, have your children go to the local schools and become part of the community."

"That makes sense, but where in Europe can we have that experience?" I asked.

I decided we should dip our toes into the water, so I gave Nancy plane tickets for Christmas for a late winter trip to France. Neither of us had ever been to France and my thought was business is lousy, let's go to see if we feel the joie de vivre and have some fun. In March of 1992, Nancy and I flew to Paris.

Late winter is the time to go to Paris. There are fewer tourists, the days are getting longer, the weather is mild, the light is lovely, and well, Paris is Paris. Even after thirty years and countless trips back and forth to the States, I still get excited when I arrive in the city of light. We stayed on the left bank and thought of ourselves as following in Hemingway's footsteps. We walked and ate our way through the restaurants, neighborhoods, museums, and markets that seem to be permanently in place.

Through foodie books, not Michelin, we hit many of the smaller brasseries and bistros and we had our first food lesson.

The name of one especially memorable bistro escapes me, but it was hidden on a ruelle, or alley, far north of the Gare de Lyon and had a classic zinc bar, carafes of wine, waiters in black aprons, and it was filled with locals. Young children sat patiently late into the evening without having to be entertained with their electronic babysitters. We too observed the art of eating and conversation with curiosity. As the only Americans, actually the only tourists, we happily avoided the tourist ghetto section. I played it safe and ordered a Charolais steak rare (saignent) and Nancy ordered rognons de veau. Nancy, a

French speaker, recognized veal on the menu (carte) and thought it would be a safe choice. The waiter could have and usually would have left us to our own devices, but to his credit he looked down without condescension and said: "Madame, savez-vous ce que sont les rognons de veau?" (Madame, do you know what rognons de veau are?) Nancy replied in very good French, "Oui, je les aime beaucoup." (Yes, I like them very much.)

He replied "Quel cuisson?" (How do you like them cooked?)

"Rosé si vous plait." (Pink please.)

"Très bien madame," and off he went.

This was one of those "Danger, Will Robinson" moments from the old TV series *Lost in Space* where somewhere in your subconscious something is not quite correct and faint alarm bells are going off, but you cannot quite put the signals in their proper order. Sauvignon Blanc and inexpensive Burgundian red flowed along with a dozen oysters, delicious mixed salads, then my steak with a wonderful light Burgundy sauce, frites, and Nancy's rognons, a plate of veal kidneys sliced in halves in the form of pink mushrooms sautéed in the same lovely Burgundy sauce. Now, if you have not spent time on a farm, have not eaten mountain oysters, tripe or other offal, the smell is a bit stronger than most suburbanite Americans are used to. What to do?

Nancy would agree with me that she is more of a traditional meat and pota-toes, grilled fish girl, so without missing a beat I said, "let's switch." Quite simply, I was not going to be a wimpy American and was going to eat the things no matter what. At that moment, another one of my love affairs with France began that continues today. I love les abats (organs) and especially kidneys, but I do draw a line with tripe (the second stomach of the cow).

After Paris, we headed to Burgundy for several tastings that we had arranged through Sidney. As neophytes to the rural wine region that was highly disputed throughout history, our initiation was at a high level. We descended into some of the best-known cellars and were transported with tastings at Domaine Denis Bachelet in Gevrey-Chambertin and Domaine Frederic Mugnier in Chambolle-Musigny. Context is important because at the time Denis was just beginning to be known and the 1991 vintage was only Fred's seventh. I was even able to buy an older bottle directly from Fred by simply asking, an impossibility today.

We visited Domaine Albert Morot, run by the diminutive Mademoiselle Chopin who was perpetually dressed in a black blouse and skirt and often seen riding her bicycle in Beaune. She was no more than five feet tall and ninety pounds wet, had a petite mustache and reminded me of my 6th grade Latin teacher who kept us boys in line, and ran a tight ship. Mademoiselle Chopin's Addams Family–style house and three-story deep, stone cellar stair-case is a wonder to the power and skill of masons.

Our Burgundy stay was brief. I do not even remember where we ate, but Sidney had arranged for several tastings that blew our socks off. She also encouraged us to meet and spend time with a wine broker and agent who became our raison d'etre to move to Burgundy. Becky Wasserman was a petite woman with a long mop top of curly, black-turned-grey hair. She was a chain-smoker who told it like it was. Despite her gritty New York upbringing, Becky had a way of bringing Burgundy and its vignerons to life. Sidney and Becky, I would come to find out, were kindred spirits: two women who excelled in a traditional male profession. This was 1992 and both Sidney and Becky had been in the business since the 1970s—Sidney importing and selling wine and Becky starting in the selling of French oak barrels to California wineries followed by representing and selling the wines of small Burgundy domaines. Back then, wine was an exclusive, white-man's club.

With today's efficacy in communication, it is sometimes difficult to put in perspective the role of an importer in the 1970s and 1980s when importation was built on a personal visit, a handshake, and a handwritten purchase order. Phone calls were the first high-tech method of communication between the buyer and the seller all the while in the seller's native tongue. Telexes were next followed by the fax in the late 1980s and today, email. However, importation is still a personal business, especially in a place as small as Burgundy, where life is familial and family relationships last generations.

Becky began the process of educating the world on these small wine craftsmen and women in the late 1970s by literally carrying samples from her home in the hills near Beaune to Boston, New York, Chicago, St. Louis, and San Francisco, testing buyers on the wines, taking orders, and most importantly, relentlessly promoting young and upcoming winemakers. Her career, spanning nearly fifty years, saw Burgundy develop into the world's standard for pinot noir and chardonnay in no small part due to her efforts. When Becky invited us to come up to her home in Bouilland for an apéro we were entering into a special world. Raising a glass in someone's Burgundy home is still defined by trust and an understanding that Burgundy's wines are personal, not a commodity. They are the essence of life.

Now an apéro is not just a cocktail, nor is it just hors d'oeuvres as we say. It is much, much more than that. In Burgundy, it means come over as you are and don't make any other plans. Have a glass—stay, talk, eat a bit, and see what transpires. If you want a drink, you go for petit coup at the local bar. An apéro implies effort on someone's part; not dinner, but not just come over for a beer either.

Bouilland is one of Burgundy's most magical places. It is located about ten miles from Beaune up a small valley that empties into Savigny-les-Beaune. A creek, runs down the valley from a spring near the headwall of a box canyon that surrounds Bouilland. Becky's farm, Le Serbet, is a wonderful collection of ancient buildings that she and her late husband Bart converted in the early 1970s. There was no better hostess, raconteur, and saleswoman than Becky, along with her second husband, Russell, who manned the kitchen. The warmth you felt when you entered their home was legendary. Their knowledge went deep; a repository of Burgundy and its history.

Becky's home and Bouilland are not hard to find in daylight or on your second visit. But on our first pre-GPS trip in March 1992, it was dark, and we were lost. In Savigny-les Beaune, I heard the distinctive thump-thump of a deflating tire (the first flat tire in our married life). Knowing what a special invitation this is, I am thinking, you have got to be kidding me. This might be our one chance to find a way to actually get to live in France and here we are stuck in the middle of nowhere (*le millieu de nulle part*, as I would learn). Thanks to my upbringing around parking lots in DC where flat tires are quotidien, we were barely slowed down on our quest. I can change a tire fast, even with just a poor street lamp. In seven minutes, we were on our way.

Bouilland is not what you call a metropolis. It is the definition of a French country village: a population of 130, few if any signs, multiple alleyways, streets made for tractors (wagons), and street lamps that are useful if you know where you are. After circling the village several times, about a two to three-minute exercise, we found Becky's home. We only recognized it because of the fine stone gate that Becky told me to look for. Becky greeted us graciously and we sat by a fire in her massive fireplace. In no time, she pointed me in the direction of her cellar to fetch the first bottle of champagne and our apéro began.

It is amazing what two bottles of Champagne, some Citeaux cheese from the local Cistercian Abbey, a well-stoked fire, and a cold wind blowing outside will do to the tongue. It was clear as we began our conversation that Sidney had spoken to Becky, and prepared her as one turns the soil, for our visit. Over a two-and-a-half-hour period, Nancy and I poured out our hearts about how we wanted to move to France, start a new family adventure with our children and leave the States to work in the wine industry. Becky asked, "Do you know what a stagiaire is? It is French for an apprentice." She continued: "Each year I employ one or two college graduates who come to work for a year with me, take clients on tastings, collect samples, and set up dinners for clients. Yes, my stagiaires are gophers, but they taste a lot of wine. Would you all like to come work for me?"

Incredulous that our dream seemed so easily within reach we instantly said "yes, as long as you could wait a year for us to break the news to our children and the rest of our family."

She said "why not," and the deal was sealed. Such was our first apéro.

Sometimes when deals come easy, it is easy to get skeptical. But something told me in the flow of the evening, that this time easy meant it was right.

After coming down from our Champagne and new adventure buzz, we were surprisingly sanguine about our plans. We had spent a year discussing the various aspects of the idea and had seriously examined how our children and we as a family would adapt. It was clear when we began our discussions that unless both of us were totally committed and unified about the idea, the children would sense an opening as a shark smells blood. This was the key to the whole adventure: we wanted to have an experience as a family or not at all. We would be open to possibilities that if the children were happy we might stay three to four years and if they were unhappy, our sojourn might only last one year, but what a year it would be. The children learning to speak French and us making lifelong friends in one of the world's great wine regions meant we could always come back for vacation and more importantly, buy great wine. There was no downside. The upside was perhaps we would love it, would fit in, would find new career passions, and our lives would change forever. We both believed that we would find an adventure for our family but we never believed it would happen so quickly and seamlessly.

Returning home to the States, the first thing I did was to write Becky a long thank you note with an outline of our plans to arrive the next spring in May 1993. The response was deafening. We waited anxiously for a letter, running to the mailbox for six weeks, but null. We had told no one about our departure, not even the kids, and we were almost resigned that it was not going to happen until I finally told Sidney we had not heard from Becky. "Call her" she exclaimed. "This kind of response rate is why Becky needs you."

I called Easter Sunday and Becky said: "Oh yes, I have been looking at schools for Tyler and Alexa and have been getting your housing arranged."

I hung up and said to Nancy: "I guess this means we are going!"

Now, the anxiety turned toward the reactions from our parents. Nancy's parents were excited for us even though they would miss seeing their grandchildren. They lived in the Philadelphia area and we saw them fairly frequently. They said they looked forward to visiting and discovering France and Burgundy with us.

Giving my parents the news that we were moving to France in a year was more complex. My parents, who separated in 1984 but never divorced, had a complex, albeit cordial relationship. Mom and her friends would have dinner with Dad every week or so, and holidays and grandchildren's birthdays were spent together. In all candor, they were much more pleasant to be around than when they lived together.

Before he had a debilitating stroke in 1990, dad went on many European biking trips and was especially fond of Beaune and Burgundy. He sensed I was cut from the same cloth as him, had a strong independent streak and that I needed to try out something new. But I was the the oldest of four siblings, I always felt I had to be the responsible one or so I felt. I dreaded telling my father that I was leaving the family business. Much to my surprise, my father's reaction was enthusiastic. He recognized in me his same stubborn, entrepreneurial streak and was proud to see me follow my passion. One of my favorite stories my dad often told was about the day he graduated from high school. He immediately got a job at the local mill: everyone's dream. After a day, he told the foreman "I am quitting and I am never ever going to work for anyone except myself!" His speech was limited but he smiled and said as best he could one word: "gooood." This was his highest form of approval.

Now for my mom, Alyce. Mom, a Northern Virginian by birth, with Irish, English, and German roots, was a Baptist in upbringing. She smoked, drank, and earned an MA in psychiatric social work from Catholic University (CU) in the early 1950s. In the '50s, very few women got undergraduate degrees, much less a master's. The combination of a Baptist upbringing and a master's degree from CU made her a strong, independent, and determined woman. During the 1930s, her extended family did not have the financial stresses of my father's side. This is not to say they were rich, far from it, but during the Depression they had steady jobs working for the Southern Railway, the Tesla of the period.

I was always very close to her, and I grew closer to my dad as we worked together. But after my parents separated and I had my own family, I learned we are all fallible, even parents are not perfect. Her reaction to my announcement that we were moving to France, as we say in the South, was a "bless his little heart" moment. With her lips pursed, she lit a cigarette and she said: "Well, that is good, dear. Tell me about your plans?" Those not from the South might need a translation. "Bless his/her little heart" is the ultimate, passive aggressive way of saying, "have you lost your mind?" In this case, her snide use of the old southern phrase meant: "You are leaving me? Who will take care of me?"

Subsequent generations have normalized the American way of moving to where a job takes you and where your interests lie. But for my parents' generation, moving to France with your children for a sabbatical was fou (crazy), out of the ordinary, and exotic. For me, it was as daunting as it was exciting.

Then I realized how many families with jobs in government, large corporations, or in the military move many times in their careers. You can get a great deal done in a year if you are organized. Exiting is a series of disengagements and engagements. Many are mirror images of each other and others are one offs. We had two principal lists: a business list and a personal list with sub lists for each. It is interesting to think about our priorities at the time: rent our house in Alexandria, Virginia, get a rental in France—easy. Not so fast, since we were going for a year we had to consider, do we rent our house furnished? Who do we rent to? What is the price, the insurance, deposits? For France, do we bring bikes, toys, pots and pans, electrical transformers, and books. Which ones? School: where and what grade? Transportation, utilities, doctors, and stereo and CDs. All of this requires time. Shipping: how, how much will it cost, and how much stuff can we bring? What are the taxes, what are the forms? Visas? How do we stay for more than three months?

Perhaps the biggest little detail was the visa question. Nancy, someone who always drove between the lines, was worried sick about how we were going to stay legally in France. No problem I said: "Once we get there, and as long as we stay under the radar, we are bringing money into the country, not asking for social services but are actually paying for them, they will say "stay as long as you like." However, worry is what spouses do and sometimes a worried spouse makes for a resourceful ally. Luckily, Nancy's grandmother was born in Ireland. In 1992, the Irish were happy to issue you a passport for a small fee as long as you could prove citizenship through the grandparent's birth certificate. As you would expect, the parish church had burned down with the records. How many Annie O'Sullivans were there from County Cork? Enter The Dubliner, an Irish pub on Capitol Hill in Washington, DC, and the FBI.

My dad's best friend and drinking buddy George O'Keefe told me: "I have a friend with the FBI who helps find the birth records of relatives in Ireland for a small contribution, say two hundred dollars that goes to help the old country." (Can you say IRA?) I called said agent, we met at The Dubliner, exchanged Nancy's and her grandmother's information and the cash. About two months later, a birth certificate arrived from Dublin. You can't make this stuff up.

Birth certificate in hand and passport in process, we made our lists, we also began to prioritize and perform a triage on what was really important and what was not. To our amazement, we found we did not need all that much stuff. We could sell our cars in Washington and find one perfectly good used car in France. We could take apart our bikes, put them together again and use them for transportation in Beaune. Beds however, are not the same size because of the metric system, so we simply bought new sheets to get by. Utilities, not so simple. It is easier to get a mortgage than it is to get utilities in your name in France. The children began French classes at their schools, I went to the community college to learn the language I never fully picked up in college and Nancy kept track of the lists. Line-by-line we inched closer to May.

In November 1992, I visited Becky in Bouilland and looked at two options for the children's school: the local public school in Savigny that corresponded to Alexa's 6th grade and for Tyler, the middle school for 8th grade in Beaune. I made the executive decision that we did not need to be driving to two different schools, ten miles each way, and I visited the local parochial school with les bonnes soeurs (the nuns). The facilities were out of a Madeline storybook with the kids (not in uniforms) but lined up quietly to enter the courtyard each morning with athletics—a ball to kick and rope to jump.

I took stock of the house where we would be living: a delightful converted bergerie (a sheep barn) on Becky's property. It had three cozy bedrooms, original 16th-century beams exposed and sheep grazing in the field outside.

I also experienced my first Hospices de Beaune Vente des Vins (wine auction) weekend. Imagine Super Bowl weekend, the parties, the press, the energy, the food, the booze, and finally the game: La Paulée (harvest party). La Paulée of Meursault is the biggest of BYOB party where, in this case, the players, the best winemakers in Meursault along with their friends (often other great winemakers), share their best bottles.

The Hospices de Beaune, also known as the Hôtel-Dieu, is still a functioning city charitable hospital. But the old building is now a museum. Built in the classic style of 15th-century Burgundian architecture with a Flemish tiled roof, it was founded in 1443 by the Chancellor of Burgundy Nicolas Rollin. Since 1859, it has also been the site of a three-day mega event, a charity auction on the third Sunday in November celebrating Burgundy's famous food and wines known as Les Trois Glorieuses (the three Glorious Days).

It begins with a black-tie dinner at the Clos de Vougeot (imagine Cinderella's ball) on day one, and ends with lunch at La Paulée de Meursault on day three. On Sunday, day two, The Domaine des Hospices de Beaune, the winery of the Hospital, which owns around 150 acres of donated vineyards, auctions off to the wine trade and public barrels (a barrel is 24 cases of wine, 288 bottles) of young wine from the current vintage. The auction, run by Christie's from 2005 through 2020 and taken over by Sotheby's in 2021, is also online, so anyone with a computer can buy a barrel from anywhere in the world. This is not to say you can take the barrel home, but that the barrel you buy will be raised (élevage), aged in barrel by one of the local wine merchants, usually for twelve-to-eighteen months, put in bottle, and then shipped to you.

Anticipating my first Trois Glorieuses made me heady, but boy did I have a great deal to learn. It was a lot just to learn the geography of the area, dotted with little towns named after the vineyards that surround them. In each region there are multiple vineyards with various expositions. As I was to learn terroir, is not just the dirt, but the geological composition of the soil in the context of its position on the slope. A bit more northeast, southwest, or west? What about the airflow from the valley above it? Cool or protected by an outcropping? This and multiple other nuances I would learn are the key to understanding Burgundy's vineyard's personalities. These personalities are expressed through the Chardonnay and Pinot Noir grapes that react and form their typicity, yes, their distinctive tastes from the above factors. Throw in the geodesic dome of the weather of a particular year, you have a taste landscape that changes from year to year.

Over the winter and spring of 1993, I joined Becky on several of her less glamorous sales trips to New York and Boston to begin to get a sense of the business as well as develop friendships that have expanded into the next generation of our sons and daughters. It was all so new and overwhelming.

The most memorable event I went to was in January, in my hometown of Washington, DC, without Becky. I was invited to attend a tasting of Domaine de la Romanée-Conti's 1990s led by Aubert de Villaine. He was the legendary, kind and thoughtful managing director of Romanée-Conti with whom Becky had been friends since her arrival in Burgundy in 1968. Domaine de la Romanée-Conti's wines are considered the pinnacle of pinot noir and are some of the most sought-after bottles on the planet. The tasting is still imprinted in my mind. As I entered the tasting room at the Four Seasons in Washington, DC, there was an energy in the room and it was coming from the wine. This energy is best described as charisma from the Greek Khárisma, which means favor freely given or gift of grace. It is more commonly thought of now as a spiritual power or personal quality that gives an individual influence or authority over large numbers of people.

This gift of grace was in the physical form, one hundred place settings each with seven dazzling glasses of wine: six red, one white. The smell of freshly crushed blueberries filled the room. It remains the most overwhelming and vivid sensory experience of my life. The seven wines were all distinct and different. From the same family, but each with their own personality.

The wine's distinction, you could feel, was going to be accomplished in their own way: there were dancers, cello sonatas, tenors, string quartets and sculptures in progress. I was in heaven. "What is going on here?" I thought. "Do others entering feel (yes, feel) the smell of the wines like I do?"

My notes on the tasting have long disappeared. But after the tasting as Aubert was speaking to guests, my nose lingered over my glasses, smelling each wine's perfume, and not wanting to leave. Why are these wines like this? How can there be energy from a glass of wine? How can they all be so different, yet so specific to such a small place? What is this place, Burgundy? What is going on there? How can these layers of smells and tastes come from grapes? I was confused, excited, and utterly hooked.

Chapter 3

Un Mer de Jaune (A Sea of Yellow)

Our departure funnel narrowed as we continued to check off our goodbyes, select books, toys, kitchen utensils, and clothes to pack and attend parties with friends. There was lots of family drama filled with tears of "we will never see you again," and too many jokes about one-way tickets.

> The final couple of weeks before we left the states were frantic to say the least with the last two days simply exhausting. I quickly learned about the docks of Baltimore and consolidation services, ocean shipping and customs in France. Ironically the French were the ones who made it a fait-accompli and moved our things to us with and in good form (except for delays because of national/ religious holidays—i.e., goof-off days) while the boys in Balmar let our things sit on the dock for two weeks before they even made the boat. If you ever go through a similar experience remember this one word—pirates. All transporters are pirates and buyer beware.

I wrote.

Departure was a Saturday and our dear friends Jane and Esko volunteered to take us to the airport. We had lunch at our home in Alexandria with my father and Joe, aka Saint Joe, who drove and took care of my father since his stroke three years earlier. We anticipated that having Joe and my father take us to the airport would be too emotional and we were correct as Dad burst into uncontrollable tears when we loaded the cars and said our goodbyes at the curb. Families are a joyful burden, both wonderful and terrible as we mature, become parents in our own right, navigate our own lives, and learn to love our parents differently. Closeness is never apparent until there is distance. The simple and often silly constructs we impose on our family lives are so much a part of us that we cannot see them until distance is imposed. We were not going to war; we were going to France for a wonderful family adventure but the departure, really the unknown, took over the moment.

Way before TSA, flying was well, classy, or at least it seemed a great deal more civilized than it is today. Flying was something special that you got

Chapter 3

Alexa and Tyler, spring 1993, Alexandria, Virginia, preparing to go to Burgundy.
Courtesy of the author

dressed up for: boys in coat and tie, girls in dresses, a hot breakfast with real scrambled eggs and sausage.

After checking in and passing through security designed for the "Take this plane to Cuba" era, the children began to get pensive and fussy as if they were homesick already. Alexa, who early on was the most interested in the adventure but became the most reluctant, sat on the floor of the terminal, back to the wall and said: "I do not want to go. It is going to be terrible. France is going to be just like Africa."

I looked at Nancy and she at me as we tried not to laugh. "It is going to be fine. You are going to make so many new friends." We placated her as we kept looking impatiently for the signal to board the plane.

Once we were on the plane the kids' anxiety subsided. It is interesting to reflect that the children were more focused on the size, condition, and food on the plane than on the trip or eventually living in France. Children focus on unknowns that they can grasp not on those that are abstract such as living in a new place or foreign country. Adults on the other hand must remain solid and never show fear of the unknown in order not to undermine the family's confidence for the task at hand. This is natural but we couldn't stop wondering where are we or why did we come to France?

The concept of refection, reflichir in French, would come to dominate our lives. Reflection, nuanced colors, no black and whites, more greys and pastels are the French way. It is clear to me now that in the three decades since I arrived in France, I fought the "French way" many times but learned to embrace La France, its people, and the pleasure of nuance. I found in my friendships, my interaction with strangers, my ability to "get things done," that understanding and embracing nuance is a much richer and more interesting way to live. It requires patience, a patience most Americans do not have.

We came to France to experience something new. This sense of discovery, discovering a new language, friends, customs, food, a way of living that intrigued us: we were not afraid to experience new possibilities in our lives and we were excited to create a new life for our family. If we stayed in France permanently, it would be a radically new life as compared to the States. Even if we stayed just a year, all of us would have an imprimatur that would make each of us different in our own unique way. There was no downside.

To Air France's credit, their flight crews were terrific, the attendants delightful as they spoke to the children in French: "Ça va le repas? (Was your meal OK?), "Peux-je voir votre dessin?" (Can I see what you have drawn?), "Aimez-vous le film? (Do you like the movie?) Blessedly, after dinner the children passed out and did not awaken until breakfast was served. Our arrival in Paris at Charles De Gaulle Airport was exciting with the promise that the kids would see hundreds of bunnies in the grassy areas between the runways pocked with holes. The bunnies making their home as prairie dogs on a prairie, did not disappoint.

A daze is normal after a night flight to France from the East Coast. It is always difficult no matter how many times you have done it and especially so after the drama of our departure and two young children in tow. Collecting our multiple bags as modern day "Okies going east" we met Jo Thorton, Becky's apprentice whom I was replacing. Jo, a tall, lanky, permanently disheveled and adorable young Englishman, carried himself in the manner of a prep school educated boy. With a bit of Harry Potter like magic, he stuffed the Renault Espace minivan full of our bags and headed south to Burgundy (Jo later became managing director of Moet Hennessey, Europe).

There is probably no more beautiful time to drive south in France than in April and May. Spring wakes up with the smell of lilacs, wisteria, and cherry trees making the air as fresh as it is after a nor'easter. The hoary frost, grey, fog and humidity of the winter are forgotten as the sun rolls north at different paces. In a country the size of France, a few hours' drive leads you through many climates, thus the months of April and May can be precocious, tardive, wonderful, and malicious.

May brings rapeseed (colza), the most perfect yellow that exists. A riot of yellow flowers that coats tens of thousands of hectares from Paris to

Burgundy. My first memory of France is renewed each spring and brings a smile to my face. I remember my family asleep in the car as we sailed effortlessly south through a sea of yellow.

PART 2

Tourner le Sol

Turning the Soil

We turn the soil and feed it with composts. The vines are our children. We must protect and nourish them so they will bear fruit. The plowing of the soil in the fall allows us to easily comb it back in place in the spring as part of the long-term nature of farming, especially viticulture. We trellis, pull up dead vines for spring replanting, prune. Tending to our vineyards and their vines is a husbandry of a most subtle kind.

Chapter 4

Quel Grand Pays
(What a Great Country)

Joyeux enfants de la Bourgogne
Je n'ai jamais eu de guignon
Quand je vois rougir ma trogne,
Je suis fier d'etre Bourguignon
Je suis fier, Je suis fier, Je suis fier d'être Bourguignon
Je suis fier, Je suis fier, Je suis fier d'être bourguignon

Happy children of Burgundy
I never have had bad luck
When I see my red face (from drinking)
I am proud to be a Burgundian!
I am proud, I am proud, I am proud to be a Burgundian! (refrain)

The traditional Burgundy drinking song "Joyeux Enfants," along with the "Le Ban Bourguignon" or the "La La" song (a ritual of singing la la while waving like the late Queen Elizabeth and clapping hands) are Burgundy's drinking songs. At any event in Burgundy—dinner, drinks with friends and in bars—these songs will spontaneously erupt early and often. We have no similar drinking song in America that I can compare them to. It is practically religious in fervor and shares an unspoken collective ethos as "l'hymne de la Bourgogne." They evoke the same shared passions we have for sports team or university fight songs. Think of the Michigan, USC, or Notre Dame fight songs and you get a better sense of the depth of feelings and shared history that only the chosen can truly experience.

The song is ubiquitous to La Paulée of Meursault and Chevalier Taste du Vin dinners (think of Fred Flintstone and the Royal Order of Water Buffaloes and you are on the right track). It is accompanied by normally well-behaved and chicly dressed women and men in black tie standing on their chairs using their voices to the limits of their lung capacity and twirling napkins over their heads. These are the happy results of several good glasses (or more likely bottles) of Burgundy.

One of the key issues we faced when we arrived in France was how to navigate schools for the children. Everyone said that it will work out and that was our family line, but to be honest, even after reading ex-pat books, we were always a bit apprehensive. We were committed to living a non-ex-pat existence by sending the children to French schools, having French friends and living a life of participation, not observation. This was of course easy being in Burgundy because at the time, we were the only American family.

I'm sure we felt the same consternation that families moving to the States felt when the kids do not speak the language. The initial shock for all children is bewilderment and fear, but thrown into the deep end, yes immersed, children thrive with a bit of patience and understanding from their parents and teachers. I think of all the different waves of immigrants that have arrived in the US and their struggle to assimilate. The children end up speaking the local language flawlessly. After a year, they roll their eyes at the struggles and butchered accents of their parents.

We arrived in Burgundy toward the end of the children's school year on purpose so they could finish the year in the French system. The logic of placing our children directly in French schools was to get them exposed to the language, make friends, not worry about grades and be in a good place both academically and socially in the fall. Let's not fail to mention the need to get them out of the house and not to be under our feet all day as we adjusted to our new home and work.

Tyler and Alexa went to the private Catholic school, St. Coeur in Beaune. Do not imagine the tuition was a heavy burden. I have the invoices, $110 per year per child with lunch additional. *La cantine* (pronounced conn-teen) cost $382 per year. Immediately, I decided I like this country's priorities. After the first week, the kids said lunch was delicious and included fresh vegetables, a salad, warm baguette, fresh juice, a starch, meat or fish, and dessert. The kitchen chef, Chef Anne Marie, not "The Lunch Lady," always said "bon appétit!" (although the kids to this day shudder at the memory of her revealing her hairy armpits as she served the food).

The school day is a long one in France starting at 8:00 to 9:00 am, depending on the grade and finishing at 5:00 pm. The full day is broken up by a two-hour lunch break with many of the kids who live in the neighborhood going home for a family lunch. The first year living in the village of Bouilland about ten miles from Beaune, we were relieved of this duty by Chef Anne Marie. For the next three years when we lived in the center of Beaune, it was a delight to have the kids walk home for lunch, often with their friends, and break bread together. This is still evident throughout France but has unfortunately declined as both spouses work. In smaller towns (provins), generally anywhere outside of Paris, this tradition continues due in no small part to the extended family of grandparents who often are present and retired (en retraite).

Village life is one that everyone should experience and one that I still embrace because of its inclusion. There is involvement by everyone nearby; this is not to say busybody neighbors, but a familial concern for others that creates a community (commune). It takes time to become part of this extended family, but eventually we became sort of cousins, the French expression for extended family. That fall, Tyler had to apologize to Mayor Robert for pulling down some political posters as a prank. Robert kissing Tyler on the forehead and thanking him for his admission and honesty was a wonderful moment of character building. We participated in the fall "Fete de la Chasse" Fall Hunt Festival with the sanglier (wild boar) prepared by the 2-star Chef Jean-Pierre Silva, dancing with the locals to accordion music, and spinning not just from the dancing but from the plentiful red and white Burgundy.

Wednesday afternoon was reserved for sports, pony club, music lessons, and other outside interests (loisirs). Sports as a sixth class, or as the only class in many US schools, is not the norm in France and is reserved for discrete moments. It was a relief not having to be a bloody soccer mom or dad running all over creation trying to enrich them (jam extra activities in our children's lives with the vague hope of getting into a better college).

A memorable Wednesday afternoon that first fall was when Nancy's parents visited and her dad took Tyler to soccer practice. On the way, Pop was pulled over by the gendarmes (police who need no probable cause) and asked to show his papers, which of course he did not have (tantamount to a felony). Tyler explained in perfect French that his grandpère from the States, did not speak any French but was taking him to soccer practice and he hoped that they would understand. Tyler also assured the gendarmes that he would make certain his grandfather had his proper papers with him for the return (retour).

As the fall progressed, the children were learning math and writing French poetry and songs. It amazed me to see how quickly they became proficient. To this day, I cannot come close to duplicating their proficiency. There were poems about La Vigne (the vine), Chanson D'automne (song of the fall), and folk songs specific to Burgundy. In the fall, Tyler learned part of Les Vendanges, a 1858 poem by Victor de Laprade:

> *Les Vendanges*
> *Avec des cris joyeux ils entrent dans la vigne;*
> *Chacun, dans le sillon que le maître désigne,*
> *Serpe en main, sous l'arbuste a posé son panier.*
> *Honte à qui reste en route et finit le dernier!*

The Harvest
With joyful cries they enter the vineyard;
Each, in the row that the master designates,
Scissors in hand, under the vine laid his basket.
Shame on whom tarries and finishes last!

That first November, I asked Tyler the great American question: "What did you learn in school today?" He slumped his shoulders and said, "some Burgundy songs" and I asked to see them. What he showed me was more precious than a lost copy of the Declaration of Independence, the Magna Carta, or even the original handwritten lyrics to "American Pie," "Freebird," or "Born to Run." Here was a copy of "Le Bourgogne," the Burgundy drinking song. What other proud and noble race would teach 6th graders drinking songs? I thought to myself, yes, this is a great country, this is where I want to be and, yes, I am proud to be a Burgundian.

La la la la lalalalalère
Lalala Lalala lala lala

Chapter 5

L'Ecole et les Vacances (School and Vacations)

The French school system is series of paradoxes. The fact that the children are not forced to fill every moment outside of school to better themselves is the simple reason that at the end of the day, they are spent. The French school experience is not a warm and fuzzy environment, but a one-size-fits-all approach. (There have been changes since my children were in school but they have been incremental, not philosophical differences.) Simply put, there are no participation prizes in France, yet when kids are not in class, they are left alone to be kids. Over the years in speaking to other ex-pat parents, we shared many of the same stories: there is something to be said for being home in the evening and not to worrying about where the next sports practice or game is going to be.

At first, Alexa and Tyler struggled with school. Bless Becky, her husband Russell, and others in the office for helping them with their homework and the dreaded dictée (dictation of spoken French) that first year. After school, the kids would set up shop in Becky's office in Bouilland and do their homework (devoirs) with the French team doting on them to help with their spelling, pronunciation, and how to navigate the system. Tyler and Alexa made amazing progress. In eight or nine weeks, they were communicating and within five to six months were fluent: ah, to be 8 and 11 again! Their grades (notes) went from average to very good, and as I look back at their teacher's comments on their report cards, I am struck by the sincere interest the teachers had in their progress.

The French school system is very rigorous and wonderful for the upper-level student, but if you are a slow learner, bored, or are a typical 13-year-old boy and totally in another world, God help you. This rigor even extends to the list of supplies each student is expected to have with them on the first day of class and is an annual trek of marathon dimensions. The list is detailed and mind blowing from the type of colored pencils down to the

31

specific erasers you should have in your pencil case (trousse). The trousse becomes a child's obsession and a parent's nightmare. In 1993, there were few office supply stores so we were relegated to buying everything from the local paper store that for some reason was the only place you could find the stuff. Coincidence?

This standardization, or system based on equality, is of course blatantly unequal based on how children learn. The system assumes everyone learns the same way. I shudder to think how at least 30 to 40 percent of my high school classmates, all boys, would have fared, least of all me. My children were good students (they got this from their mother). When they returned to US schools four years later, they were ahead in some classes, behind a bit in others, but for the most part, I have nothing but good things to say for what the French schools did for Tyler and Alexa. In essence, the system is terrific at the pre-K to middle school level, but as kids enter 8th grade, a rigorous tri-age and tracking begins (it actually began from pre-K) where the students are culled and sent to high schools of various rigor and expectations.

Grades correspond roughly to the US system of pre-K (primaire école), middle/junior high (collège) and high school (lycée) except the class designations are reversed with kids finishing the lycée, or premaire to be followed with another year, Terminale, that prepares you for the Baccalauréat, the dreaded Bac.

It is impossible to convey the stress, fear, and loathing induced by Le Bac. It is a series of tests over a week in June that sees how much you retained over the last twelve years and will determine if you go to and at which university you will be placed; basically, your life is determined by the results. From late May through June, the nightly news has stories of families, yes, the entire family gets involved, helping the student get ready for the test. Tutors, extra classes, adults tearfully recounting their Bac experiences with hopes and fears of advancement tied up in the results of their stressed out 18-year-old. We see tests being printed, sealed in special boxes, delivered to the schools, and then opened and controlled under strict supervision on TV (a security that far surpasses those of US ballot boxes).

This is serious business. Our SAT tests are child's play by comparison. My first job after graduating from the University of North Carolina was as an 8th-grade English teacher. So, I took a special interest in the Bac and actually looked up the details to understand the visceral feelings families face. I challenge you to do the same. The Bac is a very scary series of tests on philosophy, math, science, history, French, and foreign languages. After reading about the Bac, I now know why French kids start smoking and drinking at such an early age. It is also to a large measure not standardized, has a series of complex written essays to answer, all graded individually (talk about the luck of the draw and the need for good penmanship). Finally, and here is the point, your grade directly relates to what university you will enter, period. No legacies, no extracurricular, no sports, just cold hard numbers with every parent's dream of their child scoring high enough to enter one of France's elite schools. (Changes to the BAC have occurred and my French friends now say it is so much easier for their children than it was for it was them. Your final ranking now includes your last two year's grades, but it still includes written, oral and the dreaded philosophy exam [philo].)

The French school year is based on a national standard where every student starts and ends the year at the same time, studies the same subjects and even has the same lessons on the same days. When we arrived in France in the spring of 1993, there was no such thing as a core curriculum in the US. Since then, the US has begun and certain school districts have adopted different types and standards of a core curriculum. Our fledging national standards are core light in comparison to the French system. Finally, school vacations are set in stone by the government in Paris. Winter and spring vacations are rotated from year to year in three zones to mitigate traffic jams as everyone departs and returns on the same days. It also, logically, spreads the business out over a six-week period rather than two weeks benefiting local businesses (les commerces).

The academic year corresponds to that in the States except for the more numerous and longer vacation periods. Once I added up the number of hours the French kids were in school verses a typical American year and they were virtually the same. What is not the same is the pace and the aforementioned vacations that break up the year and drive parents mad, give teachers way too much time off, and allow the bureaucrats to organize and measure results. Simply put, the kids have a two-week vacation about every six weeks, wrecking work and life schedules. Even after all these years, I still do not know how parents manage it or how we did.

This vacation cadence, really a six-week sine curve, starts in the fall with Toussaint (All Saints Day) November 1, followed by Christmas vacation. School returns after the New Year followed six weeks later by winter vacation. But be sure to know if you are in Zone A, B, or C. There is an official Education Department site that clearly states the vacation periods, the three zones and their respective departments. (Today, there is even an unofficial website that has a countdown calendar by zone to the next vacation: days, weeks, hours, minutes, seconds. I kid you not.) Next is spring vacation that falls around Easter (Paque) and finally a longer stretch of school of about eight-to-nine weeks that ends at the beginning of July.

For our first vacation, we decided to drive to Normandy and visit the battlefields and cemetery that are a must, no matter what your nationality. We managed to start after some minor repairs and a new battery, a version of a Fiat 500 that had been sitting dead in Becky's driveway. The resurrected Fiat had the pep and high-toned engine that sounded like a spinning top, and with less structural integrity than that of a Tonka truck. It was our steady companion until December when our 1991 Passat became the family car of legends. To this day, I do not know where we put the bags in the Fiat, not to mention two kids on our various excursions.

After Normandy, a swing north and a visit to the port of Honfleur left us with memories of a crowded holiday weekend with no rooms available. The holiday week of Toussaint is one of the biggest holidays of the French year. I compare it to American Thanksgiving because its focus is on the family gathering together, visiting with grandparents, and honoring family graves. We made the rounds on foot with me coaching the kids to behave and act cute but carry the countenance of a refugee to stir the paternal or maternal instincts of the various hotel owners we solicited. Tired and hungry, we finally arrived at a restaurant and hotel presided over by a statuesque woman of a certain age in a Mae West mold.

"Je n'ai pas grands choses, mais j'ai encore une grande chambre avec plusieurs lits—mais il faut que vous-mettiez les draps." (I do not have a lot to choose from, but I do have one big room with several beds—you will need to put on the sheets.)

We said great, it does not bother us to make the beds and we sulked up the tight half spiral staircase to the first room on the left with bags and sheets in hand. Yes, the room had numerous beds, including several queens and others of various sizes—all covered by rubber sheets. When we closed the doors, we dropped the sheets, laughed uncontrollably, and allowed the kids to jump up and down on the beds like 5-year-olds. Nancy and I laughed the hardest when we all realized that madame rented her rooms by the hour when needed.

Madame could not have been any nicer to us and took a liking to the children. After telling her that Tyler and Alexa were real gourmands, Madame took a keen interest in our wellbeing and directed us to a restaurant where she eats. "Il est un ami, je vais passer un coup de fil pour vous. Vous allez très bien manger et boire chez lui." (I will make a call for you to my friend. You will eat and drink well at his place.) In the end, we were so well received. All I remember is the kids asking, "When can we come back here again?"

When May arrived, the calendar in France did not prepare us working ex-patriots for its folly. It is indeed the month when everyone wishes they were born French. Imagine if our month of May had President's Day, the 4th of July, Memorial Day, and Labor Day all in one: can we say party time? If this is not enough to induce fear in employers and envy in working Americans, the words make the bridge (faire le pont) cause apoplexy or ecstasy, depending on who is issuing or receiving the check. Faire le pont (pronounced "pon") means taking an extra day off before or after the holiday that can turn a day off into multiple three- or even four-day weekends, depending on the calendar. The four holidays that fall in May into early June are May 1 (May Day), May 8 (the end of World War II), Ascension, and finally Pentecost. If these holidays (fêtes) fall on a Tuesday or Thursday, the weekend becomes a four-day affair. Throw in the thirty-five-hour workweek where many people quit at noon on Friday and you can imagine the fun.

Thus, our first May, a month where we thought we would get organized, not have to make lunches and put in some hours to learn the job became one extended vacation. We really could not get our heads around what was going on and thought, does no one work in May in France? Like most Americans, we can be a bit slow, but after the third bridge, a year when the employees won, we threw our hands up and partied on.

To say we fell into the perfect situation as a family with an 8-year-old and 11-year-old does not even begin to describe our good fortune. Bouilland in the summer is magic. Its 130 inhabitants, 30 of whom were children (32 after our arrival), allowed us to ease into *la vie belle*, a French life that was like a dream. English-speaking kids were a novelty, so Alexa and Tyler made friends instantly in the village. Literally a couple of hours after we had arrived, they disappeared. I went looking for them and found that they were playing soccer with village kids, one of whom had pulled out a dictionary. I returned to tell Nancy "This is going to work." Tyler's buddy was Julian. He became his shadow and knocked on the door every morning at 7:30 for Tyler to come out and play. Alexa made friends with Delphine who lived in the house across from a creek that became the children's summer playground along with the le tunnel de la mort (tunnel of death), under the Roman arch bridge over the creek.

I am asked again and again "how did you fit in?" "How were you accepted?" The kids made that easy. Instantly, our friends were the parents of our children's friends. Through birthday parties, Christmas, Easter, end-of-year school events, and sports barbeques, we met some of the most well-known winemakers in burgundy. We not only became lifelong friends, we became colleagues.

After our first eight months in Burgundy, we noted how life in France, especially in the country, acted as a breaking mechanism on the children growing up. It didn't halt their growth, but it was a buffer, allowing our family to take the curves a bit more slowly and with a bit more surety. In Burgundy, kids are kids for longer. The pace of the countryside, like its rolling hills, does not rush kids to grow up too fast. This pace is part of the whole system of being French that is hard to explain to non-French people. Even today, I catch myself thinking of it as the x factor that was different. The point is that it is x, plus all the other little things, that make France and the French so unique.

France is not a Chinese menu from which you can choose. It is not an either-or choice. It is all or nothing—an ecosystem that works because all the parts relate to and support each other.

With vestiges of the Middles Ages throughout Europe, France is a modern country based on a 20th-century social welfare state that was created out of the bloodbath of World War I and completed post-World War II, spreading enough wealth to keep the communist party in check (the communist party is still an active group).

This creates another paradox in that the French are fundamentally very conservative people. Not conservative or right wing in an American sense, but conservative economically. Liberal economics is seen in free, unfettered enterprise, raw capitalism if you will, from which we have social progress with winners and losers. Conservative economics sees society as an organized, slowly evolving organism that must be managed and controlled, especially capitalism.

One example is the closure of stores on Sunday. Only a few weekends before Christmas or under special local decree can stores be open on Sunday (there has been progress on this front with several grocery stores now open all day on Sunday). This also applies to sales. Yes, sales, too, are set for specific periods in January and July and called *Les Soldes*. If you are shaking your head, "What is he saying, there are specific days for sales? No free enterprise?" The answer is, and sorry Dorothy, this ain't Kansas. There is a government website that tells exactly when and in what region the sales can occur. The logic is that no store should have an unfair advantage over anyone else because of size, distribution, or, God forbid, its industriousness. This is perhaps my favorite example because it is directly rooted in the Guild System of the middle ages, where producing and pricing was strictly controlled by the guild so that no one could get ahead. The guilds forbade advertising and promotion, and it was illegal to literally stand out in front of your shop to try to attract customers!

The concept of shopping on Sunday to cook a great meal is not logical under this system. There is no need to shop on Sunday afternoons because everyone is spending time with their families. The counter argument is that when you ask the same people would you like to shop on Sunday, say go to the mall, the answer is an immediate yes because they say there is not enough time in the week to shop (grocery and large chain stores close by 8:00 pm). Or how about the argument that employees would like to work on Sunday because they get overtime and the government would get more taxes? Then unions would be against it because they are against anyone working more even though these jobs are non-union. (Unions represent less than eight percent of the work force, about five percent in the private sector, and fourteen percent in the government.) If you are confused you should be; bottom line is that most stores are still closed on Sunday and the discussion continues today.

Our family entered this experiment naively, not knowing about the existing machine that was going to shape us in so many ways. In early 1994, Nancy wrote about our first six months in France:

> The kids are truly thriving. I know it's impossible to turn back the tide of hormones, but I think we have slowed down the clock by coming to France. Tyler and Alexa are enjoying childhood to the fullest. They get plenty of homework, lots of vacation time, and every Mercredi (Wednesday) off. They are gaining useful life skills at French school. Alexa has learned to brush off the advances of Tyler's comrades. Tyler has adopted the useful phrase, "No merci!" to stifle the entreaties of older girls curious about an American date. Their handwriting has improved immensely—not Catholic yet, but definitely French. Papers no longer get wadded up in the bottom of their backpacks and math figures usually align. Alexa enjoys her Wednesday and Saturday sessions riding ponies, and she has been keen on helping out at the stables. Tyler is playing as much soccer as the weather allows. This week his dream is to become a professional soccer player. My only wish is that I had become fluent at an early age. My inhibitions become apparent when I speak in front of the kids—any gaffe elicits eyeballs-rolling-back-into-the-head response.
>
> Thanksgiving was a blur of activity and I could not have pulled off dinner for twenty-three if it had not been for Alex's brother Paul from Washington sporting his beret, and his wife, Anne, and their organizational behavior. Our two massive turkeys (the location of which was no mean feat as most French turkeys are about the size of a guinea fowl) were farmed out to ovens all over the village since ours cannot handle a large pizza. We held dinner at Becky's house so we could contain our group plus the office staff, the English wine writer Clive Coates, and visiting Burgundophiles and ex-major league baseball players Rusty Staub and Keith Hernandez. (By the way, Rusty brought a frozen turkey from New York for us!)

As Christmas approached, we listened to carols and introduced French Christmas music that the children knew by heart. There are times when most Americans think they can never get too far away from their family and sometimes Christmas is the time for distance. We were far away, but our first Christmas in France included a visit from my mother and my sister Leah. My mother, who was a Christmas carol fanatic, got off to a great start by announcing a day after her arrival "Can we listen to our kind of music?" Correct music was quickly procured and added to the mix with a slight peppering of French carols and all was well until Christmas dinner when the subject of my father's stroke and disablement rose from the table like a multi-headed Hydra.

My father's condition was really a coverup for the real question—"when are you all going to come home?"

"Your father is so feeble, who is going to take care of him, who is going to take care of the business, who is going to take care of us?" The children being quite smart, put their napkins over their mouths to suppress their giggles and ran upstairs while Nancy and I stared into space thinking, Jesus, Joseph, Mary, Mother of God, we did not get far enough away. We tried to reassure my sister and mother that dad was quite aware of his surroundings, capable of being involved in the business, had able management, and that our absence was not missed. "But who is going to take care of us?" was the refrain from my mother and sister, who was an emotional twin of my mother. They started to cry. I steeled myself and Nancy by pouring two full glasses of red Burgundy and put on more of my mother's kind of music.

Chapter 6

Les Samedis (Saturdays)

Our move to France was always couched in the knowledge that there was little to no downside. If the children are miserable, we do not like the business or the place after a year we will go back to the States having had a unique and challenging family experience. The children would know how to speak French, I would have gotten the wine bug out of my system, Nancy would have worked at her graphics profession and we would have made friends in a foreign land. If on the other hand, the children and our family thrived and we were fulfilling our professional needs, we might stay three to five years. In the beginning there was a sense that we would return to the States, but we remained open to the form the next stage would take.

And so, a year in Bouilland was followed by three in Beaune in a late 19th-century home across from the Beaune's basilica Notre Dame. The bells calling Beaune to mass, weddings, and funerals was a constant presence. The house, a late Empire style, three-story townhouse with touches of decorative lead colored windows and a courtyard, allowed us to adopt to and became a part of the rhythm of the French year. The courtyard had a cherry tree that the children would climb up to spy on the Saturday weddings across the square to the church. From May to June, they would report back on the brides, "Jolie ou moche" (pretty or ugly), the various garb the families wore and of course the hats worn by the women. French women have an incredible ability to express themselves through their fashion (mode). We could always tell when one of the wedding's family was from Paris by the flamboyance of the women's couture and especially their hats that would range from stylish to outrageously bizarre ensembles of ferns, feathers, and fruit worn with aplomb.

Saturday is market day in Beaune. Having the market literally at our doorstep spoiled us. The market begins at 8:00 am and it is best to be early. However, if you wish to socialize, or work the crowd like a politician, really the same thing, you arrive between 9:30 and 10:00. This was Nancy's routine as she would linger through the morning buying from the green grocers and specific cheese mongers: Citeaux from the monks of the Abbey de Citeaux in

Burgundy, Comté cheese from the Jura, Beaufort from the Savoie, and local goat cheese sold from the producer whose scruffy beard and long hair made him look like his milk supplier. The fish monger would pull a live fish from his tank on a trailer, bop it on the head with a billy club, clean it in thirty seconds, and voilà, fresh trout for dinner.

My Saturday morning preference was a two-hour bike ride in the hills (the Hautes Côtes) to the west of Beaune. I would always be certain to return home by noon, Alexa waiting for me impatiently for our ritual market tour.

First, we would raid the cabinets, grab several Tupperware containers and then make straight for the vendors of rotisserie chicken. Our system, actually an art, was to arrive as the crowds thinned so that the vendor might be getting anxious that he would be left with unsold birds. Alexa would look her most precious but not desperately hungry (they give nothing away when they see saliva dripping, even from a child's mouth). We would make a deal for three or four perfectly golden-brown chickens just off the spit. Most of the time, we bought the base chicken, but occasionally the chicken man was long on les grands fermiers (free-range farm chickens) and we would buy up as many as we thought we could eat over the weekend.

Countless books have been devoted to the art of the raising and roasting chicken in France and the various regulations, certifications, and rankings of this noble fowl. Yes, there is a difference in the breeds and how they are raised with the apogee le Poulet de Bresse. From a specific geographic region just north of Macon, this glorious black-footed bird, with its own appellation, is the most delicious chicken you will ever eat when roasted at 450 degrees for about forty-five minutes. Its time on this earth is regulated to have at least ten square meters, eat grains, and feed off naturally occurring insects. The gastronome Brillat-Savarin, who invented the greatest triple cream cheese, describes the Bresse chicken as "the queen of poultry, the poultry of kings."

"Est-ce voudriez-vous avoir du jus?" (Would you like some drippings?)

"Oui monsieur, nous avons nos barquettes pour ça." (Yes sir, we have our containers for that.) The chicken man, in his whites and chef hat would scoop out drippings and fill one or often two containers for us. We would hurry home, eyes down so not to have to greet anyone. On our own, Alexa and I would eat one whole chicken, steaming, dripping with flavor, and falling off the bone. Nancy and Tyler allowed us our chicken indulgence because it was our special date. We would place the containers of juice outside in the courtyard to cool (our refrigerator was micro), and wait for the fat to rise and coagulate. At dinner, we would skim the fat off, leaving the pure chicken juices to be enjoyed over the chicken and oven-roasted potatoes.

One Saturday on my bike circuit, I missed the market (and almost dinner, for that matter) due to a visit, or should I say detour, on my way home. My normal route was through a series of picturesque, stone-walled villages whose names bring a gleam to the eye of wine lovers: Savigny-les-Beaune, Beaune, Pommard, Volnay, St. Romain, Meursault, Puligny-Montrachet, Chassagne-Montrachet. My circuit varied depending on my mood or courage to mount some serious hills. I rode from the high and cool chalk white hills of St. Romain (this gives the wines their wonderful freshness), to the valley floor that narrows to the Roman road to Autun at the village of Auxey-Duresses and then through the rising fanlike valley that holds Meursault to the right and Monthélie and Volnay to the left. In the winter when the vines are bare, you can literally see the color of the soils change from pure white to a mix of vermillion, iron, clay, and marls to white again over the narrow three mile stretch of earth. The dynamic soil and the rocks that support it give the wines their range from austerity to richness to a mix of austerity and richness.

This Saturday, as I was passing through Meursault and Monthélie, my mind drifted to wine. I was in the mood to visit a winemaker friend who I thought might be home getting ready for lunch and perhaps an apéro. A glass or two of Meursault and some dried cured sausage after a thirty-five-mile ride would give me the energy to get home the last few miles and besides, it was mostly downhill. Michel Dupont, my age and a local vigneron from Monthélie, met an American from Pasadena, Leslie Fahn when she was taking her college year abroad, fell in love and married. We met at a child's birthday party and discovered that our children were in school together. Of course, there was good wine being poured at the birthday party and we began to do things as together with our families: eat, drink, and let the kids take care of themselves.

When I passed by their home, Michel was sitting on his terrace just opening a bottle of Meursault. I rode up and, as expected, he said, "Please stay for a glass."

"I really shouldn't, I am on my bike and Nancy is expecting me to make a market chicken run."

Leslie appeared and said "Stay, let me cut some saucisson, and bread, I just got home from the market" and I said, "Why not?"

Thus, began a marathon apéro as Michel served the first bottle of wine, Leslie brought out more goodies: nuts, multiple types of saucisson and then cheese. The variations in regional dried sausage is only matched by the variety of French cheeses and the particular artisanal producers or the cheese's months of affinage (aging). These products are sought after like the Holy Grail and their merits argued about amongst their faithful like devotees to a cult. As the first bottle of Meursault slipped down our throats, our tongues loosened, conversation turned to the no nos of American polite conversation—religion, politics, and sex—staples of any intelligent French conversation. The discussion, of course, included wine (and the relative merits of what we were eating) and most interestingly the details of each bottle's vintage.

As someone who is naturally curious, sessions such as this with Michel became my atelier and my classroom. Michel, as other vignerons had and would continue to do in the future, shared with me the details of what they had done correctly, where they had made mistakes and what they had learned from each vintage we tasted. I am not sure when, perhaps on a sales trip, perhaps when talking and tasting with other vignerons, it occurred to me that I could taste better than the average Joe and when I taste, I see the wine. Wine takes on its own form, color, and personality. I remember it like I remember a face, a person, a moment in time. It becomes something concrete for me.

The afternoon passed quickly as the bottles piled up, old vintages, just bottled wines, wine from cask and some bottles that were not particularly stellar. I vaguely remember several other friends coming by for a glass throughout the afternoon to give me enough good sense to go home, but decided "this is a teaching moment, I cannot leave." I also thought for a brief moment "I should call Nancy and let her know where I am, but no, she will figure I met up with some friends." At 6:00 pm, I saw we had seven dead soldiers piled high.

"I really should be heading home," I said. Michel replied "Let me drive you." It was then that I had to do some serious probability analysis:

1. Have Michel drive me home, in his current state, the same as mine, in his old, slow but solid Mercedes and I can take the risk he will stay between the lines or crash into someone or a vineyard wall.
2. Get on my bike, be wobbly, and at worst fall three feet to the pavement. I will wear my helmet and sober up during the ride home.

I chose number two.

"I was not worried," Nancy said. "I figured you were with one of the boys. Thank God you still have had enough brain cells left to ride your bike home."

Chapter 7

La Rentrée (Back to School)

As we passed our third anniversary in France, it was clear that Nancy and I had some choices to make. Alexa would be turning 11 and Tyler 14, the age where France's education system is one grand triage and kids are literally required to choose a career path. God almighty, at 14 my primary interest was football and who was playing on Monday night. My buddies and I were not dummies, but most of us were not getting 800s on our SATs. Late bloomers, forget it: no such thing in France. Throughout our stay, we felt it was important for our children to have a US high school experience and to relearn English. They were totally immersed with all of their homework and writing done in French that was clearly at a higher level than their English. Nancy was beginning to feel the desire to be closer to home and to pursue her graphic arts there, and I realized that after three years working with Becky's company, we had accomplished most of the goals we had set. It was time for a change.

Since college, Cape Cod was our summer vacation spot. Nancy and I were able to buy an old farmhouse in the late 1980s that became our respite where parents, siblings, and friends could visit. Early on, Nancy and I decided that whatever happened in France, we would no longer be Washingtonians. We sold our house in Alexandria, Virginia, and the Cape became our home when we returned stateside. After returning from vacation, we made the decision to depart France one year later and move to the Boston area where we had friends and could use the Cape house. The year remaining in France would allow us time to find schools that were suitable for the children as they approached crucial transitions: Tyler entering 9th grade and Alexa 6th. We now had to reverse course, deconstruct our French life and reconstruct a new one in Boston.

With three years in the best training ground in the world for Burgundy, I earned the dubious distinction of knowing as much as any American about Burgundy. Great, but how do you make a living with that? I did not see myself as an agent or broker. I did not really want to import wine. I could imagine owning a retail store so that I could pass along in a direct manner my passion for wine like Sidney did for me. I was at a loss. Friends in Oregon said "Come out and make wine with us," but I wondered whether I really wanted to abandon all I had learned for a new wine region and start from scratch at 40? So, for the first time in my life with many doubts and no real plan, I decided to go back to school.

In Beaune, we have the Lycée Viticole, which is the high school/junior college for wine. Many vignerons' sons and daughters, and not just from Burgundy, enter the program to learn the basics of the trade. Within the same complex is the CFPPA: Centre de Formation Professionnel et de Promotion Agricole. Roughly translated as: The Educational Center for Agricultural Professionals. It acts as the adult wine school for young adults either coming back to take over family domaines or older adults changing a career path to reclaim or reinvigorate a family domaine. It is a government-sponsored viticulture/education program with branches throughout France that focuses on all types of agriculture, such as animal husbandry, forestry, fruit growing, etc. There is nothing directly like it in the US, but if you can imagine our Department of Agriculture starting state agriculture schools organized locally like a community college, you can begin to get the idea. The major centers for wine education are in Beaune, Macon, and Champagne.

I had given lectures at the school in their wine marketing and exports classes and thus knew the program, the administrators, the teachers, and its quality. I talked to several friends at the school about the program and, as we were leaving on vacation in the States, I asked Sam if he would mind going over to the school to sign me up. He and his wife Erin, two ex-pats working in Paris, took their vacation by house sitting for us in Beaune. Sam went over to the school, did some schmoozing in much better French than mine and registered me as a student (little did I know what I was getting myself into and you all can blame Sam for what has happened in the last 25 plus years).

The program is free for French citizens because—and here is a bit of French magic—it lowers the unemployment rate. If you are unemployed or between jobs and go for a "retraining program," your tuition is paid. What's more, often you get a living stipend and you are not counted in the unemployment totals. It is amazing what happens to a percentage when you reduce the denominator. In fairness, virtually all the students were serious about the programs and most ended up in the wine business. But as a way to spur job retraining Le Département du Travail (Labor Department) gets a statistical bonus.

By and large, French education is either free or there is a nominal charge. Not being French, albeit with a work card, I did not get any of the goodies. The French are not only logical but sometimes smarter than we Americans. Their optic is, "We are happy you want to come in, invest your capital, spend your money, and not get any benefits. Heck, stay as long as you like as long as you keep spending." The tuition in 1996 to 1997 was $5,000, which was cheap especially when you consider the price of private schools. However, five grand was a lot of money for a husband and father of two with no income who is living off savings. I had resigned from Becky Wasserman after returning from vacation.

So, I made a deal with the school that I would give English lessons (mostly wine speak) twice a week. In return, I could attend classes for free and the school was thrilled because they could show another program at no cost. The class, scheduled in the early evening, was a voluntary elective and as you might expect, few showed up. By the time early winter rolled around, I was alone in my classroom. I asked myself, "Am I that bad?" and would head home to an apéro with the family.

The first week of September, there was an orientation and basic winemaking classes prior to our harvest apprenticeships. Each of us was assigned to a different winery, where we would do all the glamorous tasks of making wine: mucking stems, cleaning grape boxes, scrubbing barrels, cleaning hoses, squeegeeing the floor (raclette in French, from "racler," to scrape, as you do with cheese), punching down the grapes, and pumping over the juice.

Full-time classes resumed at the end of October. One Saturday morning after my first week of classes, the phone rang at 8:00 am. I am in bed. On the line was my buddy John Romano from Jersey. John is a force of nature; he will admit he has burned every bridge he has seen. He is one of those individuals who speaks the truth all the time, especially about the wine business but often to no good use (hence the bridges). He is rarely incorrect, but sometimes it does not matter. To put it plainly, Romano is a funny bastard. I scratched a hello in my early morning voice. At the other end of the line, I heard him bellow: "How does it feel to be unemployed?" It was 2:00 am East Coast time. He's calling me to break my balls and share his hearty, non-stop laugh. "I thought I would wish you the best and see if you are yet on the dole in France or can you get unemployment benefits simultaneously from two countries?"

"I was just thinking of you as a married 39-year-old unemployed father of two. No job prospects but in wine school in France . . . boy are you real smart! Love ya. You are going to do great. I am going to bed." Click, he hangs up.

I thought, "merde (shit), what did I do?" I am planning to move back to the States to a new city in nine months with no job prospects. I am in wine school and I am going to make a living in the wine trade? Who am I kidding? What am I thinking? In a lighthearted way, Romano gave me the focus I needed as I went back to school. I began to reflect on my new profession (métier), I felt certain that the rigor of the wine program would expose me to a unique body of knowledge that would give me many more options for a career in wine than

I had before I decided to enroll. This was not at all clear in the fall of 1996, but I had confidence that an answer would develop.

The Burgundy wine school is not strictly speaking an oenology program where you are being trained as a biochemist (really a wine doctor) with the ability to make or fix hundreds of thousands of liters of wine. The program is also not at all like UC Davis which is comparable to the University of Burgundy or the University of Bordeaux, where you get an in-depth education in the chemistry of wine, agriculture, or horticulture. Regional programs such as in Beaune, Macon, and Champagne are creating generalists (general managers, really), who can run and manage a typical French, often family, wine operation. I am half serious when I say we have just enough knowledge to be dangerous. We learn enough chemistry, botany, soil chemistry, and even driving a tractor to know who to ask for an in-depth analysis or help if needed. Specialists we are not, nor do we want or need to be.

I was able to opt out of some of the basic courses. I received credits from my days at the University of North Carolina such as math and, if you believe it, French. The program's courses in accounting, finance, management and tasting wine were easy enough for me that I often helped my classmates. On the other side of the ledger were the science classes: oenology, chemistry, biology, botany, soil geology, and its chemistry, organic farming, and pruning. That's where, even now, I tremble. Science was hard for me during my youth and taught in my native language. But as a 39-year-old going back to school in French . . . my buddy Chris from Boston said it best: "Are you kidding me?"

If I had to distill my year in wine school down to three bullet points they would be:

- Going back to school after twenty years of work is hard, especially when you are a schmoozer (aka a liberal arts guy and not a scientist)
- It is even harder in a second language
- It is harder still when you have to learn a third language: science, and speak it with new vocabulary in French

It required high energy on my part to get through wine school. I understood wine speak French but when I entered the school I had to listen to lectures that began to go very fast. I wrote notes in a combo of English, French, and Franglais and the moment I thought I am getting it, the professor used slang (argot) that I had never heard of. Everyone laughs; I am lost; he continues apace. There were times when I burst out: "Stop, what did he mean?" with my hands forming a timeout T. My classmates would double over in laughter seeing my most plaintive face. When they came up for air, they would look to the teacher and say please (s'il vous plâit), explain the concept. Other times,

I'd turn to my friend Patrick who had spent nearly ten years in Canada and spoke flawless English to give me the practical meaning. All I have to say is, "merci, Patrick," and thank you to the teachers for putting up with me.

Our class began with thirty students and ended with twenty-five, an attrition that was due to two pregnancies, one financial issue, one drug-related dropout, and one who was asked to leave because he never showed up for class. There were about twenty men and ten women with an average age of 24 years old. Three of us were not French. These days, the school is much more diverse. At least half the classes are women, a good 25 percent non-French, and the median age is closer to 30.

A curiosity of the classes—and at the time I was not sure if it was peculiar to our group or endemic to French higher education—is that everyone talks during class. The chatter is not in whispers, rather a dull roar. Imagine this in the US? You can't. To be sure, there were some jokers (losers) in the class who were taking the class to make work and then return and to run their family domaines (a scary thought). But beyond these few, the banter was constant and drove me nuts. Finally, after a few weeks of listening to what I interpreted as a lack of respect for the professors and other classmates, I piped up and shouted "tais gueule," which means "shut up." The translation is really "close your dirty dog's mouth." This is not polite French; in fact, it is quite vulgar, a bit like saying "fuck you." I stood up, walked across to the principal miscreants and actually got in their face: "Si vous voulez parler, quittez la classe. Quelques-uns d'entre nous veulent apprendre et doivent, en fait réussir ces cours et trouver un emploi. Nous n'avons pas tous quinze hectares pour prendre la relève de leurs parents!" (If you want to talk, leave the class. A few of us want to learn and in fact, need to pass these classes and get a job. Not all of us have fifteen hectares to take over from our parents!) The message was heard loud and clear although at times I still had to act as the class gendarme.

It is a given in Burgundy that when you make wine you have worked for other winemakers and have served your stages (pronounced "staj") or internships. In Europe, stages derive from the old system of learning from a maître, or master of the trade. It is a bit more sophisticated today, but the system's roots are imbedded in this guildlike system. Most of the current winemakers who went to school in Beaune or in Dijon would say their most valuable lessons were gained from an older experienced vigneron. I cannot stress enough how important this is to the fabric of Burgundy. For sure, the new generation has ventured outside the region—to California, New Zealand, Australia, Italy, South America—to learn and process different systems. In return, we welcome those from outside to mindshare. The best winemakers have had learning curves dominated by great teachers who showed them where the potholes were. This concept of a shared endeavor, where you can call a colleague and pose a viticultural or oenological question, is another part of Burgundy's fabric that was invaluable to my success.

While at wine school, I was lucky to be able to call on friends who happened to be talented winemakers for my stages. They cut me no slack, like I did to their sons and daughters when I stood up in class. At Patrice Rion's, in Premeaux-Prissey, during the harvest of 1996, my job was to pitchfork the discarded grape stems into a wagon. Then I could move up to enter the winery (cuverie) to clean hoses, pumps, hose down and squeegee the floor, then maybe do some grape punch downs before the next load of grapes came in and back to the pitchfork. All very romantic and glamorous this winemaking thing.

I spent February 1997 at Domaine d'Arlot, also in Premeaux-Prissey with Jean-Pierre De Smet the managing director, an urbane former financial type, who was much happier in blue jeans and holding court at the lunch table, where his various aprentis and clients would gather to talk about vineyards, vinification, and great bottles. Jean-Pierre de Smet, about ten years my senior, had me arranging full barrels via a poulain, a fifteen-foot wooden ladder like ramp that allows you to roll barrels up to the second or even third barrel level and then manhandle them in place with a crow bar that has a flat head about the size of a wooded spatula called un pincer (tweezers). Along with racking barrels, during which I learned the hard way to make sure the pressure was correctly adjusted, I assembled wine for bottling and counted bottles for inventory (much harder than you would think when the bottles are stacked back to front as much as three to four deep). After bottling, labeling, and preparing orders, Jean-Pierre taught me to be a fanatic about details and the crucial importance of the back of the house.

In the spring of 1997 at Domaine des Comtes Lafon, I learned how to rack white wine, the process of pumping it out of a barrel to prepare it for bottling. All of this work was not in the cellars because I was also in the vines trailing a tractor while guiding a plow by hand between the vines as well as replanting a vineyard. Note that the lowest of the low—really a terrible job—is pulling up the roots of vines from recently cleared vineyards. After the vines are pulled out of a vineyard by a bulldozer with a type of elephant noselike pick (the dexterity of the operator is amazing), the soil is then deeply tilled by a tractor and finally a team is sent in to play a version of pick-up sticks. A group of five-to-seven workers are sent into the vineyard, forming parallel meter lines—like police searching for evidence—with each worker then digging out large and small roots that are invariably imbedded in thick soil or stones as you sink in deeply with each step. This was my training ground. I loved it. After ten hours of field work, I slept well and never again would complain about the price of a vigneron's bottle.

My apprenticeships, classes, and exams progressed through the year and my French did get better, although I continued to go home at night and rewrite my notes in English with the corresponding new word in French. I think I made it clear there was never any danger of me getting into medical school, but even the basic ideas of chemical structures—stuff we read about in Tuesday's *New York Times* science section as normal and practical—was for me a stretch. Consider the word *valence*—no, not the city and its 3-star Michelin restaurant, Pic. I remember this example to this day because after a full hour of the professor talking about valences (I had not yet gotten aggressive in my time outs), I was still stumped. That night, I looked it up: the number of electrons orbiting an atom. This one word and many more like them were the new language I had to learn: the new foundation that I had to build. I was never going to have to build complex chemical chains, but I was expected to understand why they were such. This information is crucial to winemaking, but it was a slog.

The opposite extreme was biology class. Back in 10th-grade biology, I was pretty good. I asked myself how hard can dissecting frogs be today? Little did I know that in those twenty-five years the teaching opened up to include DNA and all that fun stuff about genes. To say I was in a fog as the teacher showed electron microscope photos of genes lined up, and then more chemical formulas (I thought this was biology) would be an understatement. I entered the final exam with a great deal of misplaced optimism and failed miserably: a genome researcher I am not. If you recall from the movie *Julie and Julia*, Julia Child is beside herself when she fails the written part of her cooking exam. Her French partner, Simone "Simca" Beck informs her she can retake the exam. In fact, it is her right. I found out the same and passed on the second round. (This right to retake exams remains for me one of France's greatest mysteries and one that I was happy to take advantage of and not question its origins.) Another is the public nature of grades. I approached with dread my chemistry exam. My classmates waited as the teacher graded my exam first. She walked out with a big smile and thumbs up and my fellows gave me a big cheer.

As we entered the final stretch, one of our last required wine classes was "Les Maladies du Vin," roughly Wine Sicknesses. This is what doctors would call a clinical class; applying what you have learned. As we went through the various combinations of disasters: stuck fermentations (unfermented sugar)—that can lead to volatile acidity—that means your wine is close to being vinegar, or wines that are oxidized (too much oxygen) or wines that are reduced (too little oxygen) . . . the list is endless, my classmates and I became more and more depressed. I thought, I have been in school a year and now at the end they tell us everything that can go wrong. It is so much easier simply to buy a finished bottle and not worry about it. This is way more than I want to know (I had the same dread when Nancy and I took Lamaze childbirth classes. It seemed like such a good, gentle idea until the teacher tells you everything that can go wrong).

Over a period of days, our chief winemaking professor, Jean-Pierre Charlot—big in girth, never a scowl, serious but generous, the image of a Burgundian if you ever met one and a great winemaker—patiently explained the various bad combinations that can happen to a wine. His cross-referencing flow chart, ten across and ten down that referenced these bad outcomes deflated our graduating class more and more. We thought, this is impossible, have we learned anything? Jean-Pierre sensing our frustrations said: "Stop. Let's make this simple. What we are talking about here is trying to fix something that has gone bad. If you do what you have been taught you should never have to worry about these combinations. What are the most important things we have learned this year?"

1. A healthy récolte/harvest
2. A clean cuverie/winery
3. Lots of water to clean all equipment at the end of the day without chemicals
4. Clean barrels
5. Proper élevage/barrel aging
6. Proper and timely bottling

He finished, looked at us, shrugged his shoulders and said: "That's it. You know the details. Not one of these six is difficult as long as you follow the thirty or so other steps behind each of the five above." In other words, the devil is in having the discipline to care about and follow the details all the time: "Pas de raccourcis, jamais de raccourcis." (No shortcuts, never any shortcuts.)

Somewhat relieved but still not totally convinced, we began to realize that the bad wine class was not about fixing the problem but prevention. When do hospitals have problems? When there is a breakdown in hygiene, usually as simple as not washing hands. When does a restaurant have a problem? Ditto. What separates an average restaurant from a great restaurant? Delving into the details. Of course, there is talent and ingredients, but hot food must come out hot, cold food must remain cold. Ingredients must be kept at the correct temperature under sanitary conditions until preparation. I am not saying that a restaurant at full speed is a clean place, trust me, neither is a winery. But it is a machine that works within concrete guidelines to produce sound and pleasing products. A great restaurant after service will be clean, as will our cuverie, until it is attacked the next morning for the next service or next grape arrival. Yes, success is in the details, not hard to learn but hard to execute every single day.

PART 3

Le Débourement

Bud Break

Débourement, bud break in the vines, starts in late March and is the first visible sign of spring's arrival. In the fall, the sap returns to the trunk and roots to better protect it from freezing temperatures. Milder nights wake up the sap and as it begins to rise, the buds begin to tear.

Chapter 8

Les Lundis en France
(Mondays in France)

The seed was planted for me to begin making wine in France in the winter of 1997. It was also when Nancy and I began to plan for our departure the following summer.

Jean-Pierre De Smet at Domaine D'Arlot helped me put pen to paper to plan a small, high-end winery. My objective was to create a winery (maison du vin) that was focused on buying and making wine from the highest-quality grapes. A small is beautiful approach that was unique at the time. Jean-Pierre fell in love with Burgundy and its wines while working multiple harvests. He became my mentor with whom I discussed ideas, structures, spreadsheets, costs, sales projections and above all, quality. He kept repeating "Qualité, qualité, qualité, vous réussirez tant que vous rechererez la qualité." (Quality, quality, quality, you will be successful as long as you pursue quality.)

Grape brokers were also part of my planning as they offered me wines to buy that in retrospect were extraordinary. A broker and dear friend Jean-Phillipe Lefils presented me with an offer of twenty barrels of fine to very fine wine in January 1997. Twenty years later, it was virtually impossible to find one barrel (or the equivalent in grapes) of any of these wines much less twenty. The prices were a fraction of today's prices, with the Vosne-Romanée then 5.33 euros per bottle, while the cost in 2022 was more than 25.00 euros, if you could find it. And this is the cost of just the grapes-per-bottle before aging the wine for eighteen to twenty-four months, bottling it, and then trying to sell it. This includes no production cost, overhead, profit, etc. You can come to your own conclusion—are wines today too expensive, or were they too cheap in 1996? Somewhere in between is the real answer. Fundamentally, I believe the prices were far too low in 1996 for both qualitative and demand reasons and today they are too high because there is too much demand and simply far too little wine.

When I passed my final exams in June, a minor miracle I freely admit, I was simultaneously doing the paperwork for a wine company and deconstructing our French life so we could reconstruct in Boston. Yes, I was planning to bifurcate my life with my family in Boston and my business in Burgundy. Toys that the kids had outgrown were passed on to friends' younger siblings who are are now adults making wine with their father and mother (Alexa gave her Barbies and her Barbie boat to Chantal and Frederic Lafarge's girls who to this day smile at the memory).

As I was combing through old files, family letters, and photos for these pages, I found the children's admission applications for US schools from February 1997. The task of getting them into schools proved to be much more complicated than we had anticipated. But having two polite, poised children who were fluent in French, along with a great deal of pleading and panicky calls from their parents completed the task. One example of the differences and challenges we faced was not only getting the children tested on short notice but also interpreting the results. Alexa did not do well on the math test but she was a whiz at math in France. We wondered, why? In our analysis cultural differences, nomenclature, and different emphasis in the respective programs were the deficit. In division, for example, the positions of the divisor and of the dividend are reversed, commas are used as decimal points and spaces are used between thousands. Alexa was already deep into geometry and fractions but had not yet been taught averaging and rounding. Finally, there was the question of spelling. Thank God for phonetics because in reading the papers and essays, they wrote what they sounded out in English after four years of reading and writing in French.

In Alexa's application, I found one of the written questions for admission to 6th grade (the spelling and punctuation are unchanged). The question was: Imagine that you are one of the following. Write a description of an incident in your life as:

a. a piece of clothing
b. an animal
c. a chair in an international leader's office

My family was moving to France for the winter. It has taken us a long time to decide wether to move south or go est We decided to move est. My family's name is mockingbird, we are all birds. To day is the day we fly to France My dad said that there are not too many mockingbirds that fly over the ocean, but that we could make it and we did. It ends up we lived there for four years. It was a great experience and like dad said, there aren't that many birds who can fly across an ocean.

4. Imagine that you are one of the following. Write a description of an incident in your life as:
   a. a piece of clothing
   b. an animal
   c. a chair in an international leader's office

Alexa

My family was moving to France for the winter. It had taken us a long time to decide wether to move south or go est. We deided to move est. My family's name is mockingbird, we are all birds. To day is the day that we fly to France. My dad said that there are not too many mocking birds that fly over the ocean, but that we could make it and we did. It ends up that we lived there for four years. It was a great experience and like dad said, there arn't that many birds who can fly across an ocean.

Signature   *Alexandra D Yamba*

Alexa's Mockingbird essay.
Courtesy of the author

The movers arrived in late July. They worked with astonishing efficiency, or was it more that they were moving too fast for me? Nancy directed them all the while trying to keep the kids out of the way. They seemed to have more friends than ever running in and out of the house. I huddled in the bedroom at my makeshift sawhorse desk sending off a last few emails that were just becoming normal, and I became more and more sad (triste) as the rugs were rolled up beneath me, furniture and pictures wrapped in bubble wrap. By 1:00 pm, the house was empty, the container was filled and off to Le Havre to ship to Boston.

Truth be told, I did not want to leave. Intellectually, I knew that moving back was the correct decision for the children. We wanted them to have a US high school experience and then be able to enter US universities. I also knew that Nancy wanted (and should be able) to advance her career in graphic design in America, but I really hated to leave. My friends were here in France. My passion was in France, and I saw a new adventure forming that would have to be shared with the States.

How could I make wine if I was the States? Veronique Drouhin and the Drouhin family's Oregon winery provided me the model. The Drouhins and Veronique had become great friends and as my business plan developed, I spoke to Veronique and asked her how she not only managed the Oregon winemaking, but most importantly her family life. She shared that she went to Oregon three to four times a year for crucial periods, such as harvest, bottling, blending, etc., and when not there, her team sent samples via FedEx to her. I thought I could do the reverse: sell the wine and raise money in the States with Boston as my base and FedEx myself to France to make the wine.

As crazy as it sounds, it actually made a great deal of sense because Boston to Paris was a quick overnight flight and the flights back then were cheap. In the first two years, I budgeted $5,000 per year for transportation, always flying in the back of the bus. Tickets were often just $350 to $400 including taxes. Yes, the high-season summer trip could cost $1,000 but let's be honest, how many people were on the flight from Boston to Paris in January? Thus, my commute to France about every six to eight weeks took the form of a week to ten days while trying to never be away from home two full weekends except at harvest. It worked from a business standpoint, but the family disconnect at home was difficult. When I was home, I was Mr. Mom, but when I'd been away and returned to Boston it was "who is this coming back into our unit of three?"

We spent our last night in Beaune with our friends the Danchins. Our families had become very tight since Tyler became best friends with their son Romain. They were inseparable. Tears flowed. "We will never see each other again" the ladies wailed dramatically as we waved goodbye on our way to Paris.

Monday, anticipated for years, was the day we would visit Monet's home and gardens at Giverny on the way from a final swing through Brittany and Normandy to Paris. The gardens are legendary and the French keep them in the form as Monet lived them. We were leaving the next day, so we decided this would be our last real French tourist stop. The children, after four years in France, were cognizant of Monet's imprint on the world and Nancy's love of gardens and Monet's art made it a special visit. We left Normandy at a leisurely pace anticipating an afternoon tour and then a last meal in Paris before our Tuesday morning departure. Our white 1991 VW Passat diesel station wagon (with a choke) had served us well and would continue to serve me well in the years to come. As we arrived at the gardens, a bit after 2:00 pm, the parking lot was empty and the security gates shut tight. The Griswolds had arrived at the French version of Wally World. Giverny was closed: always closed on Mondays! *Between curses, self-pity, and general malaise, we began to laugh at ourselves as we thought have we not learned anything in four years? The bloody museums are always closed on Monday.

To provide a salve to our museum misadventure (in the spirit of Ralphie's dad in *A Christmas Story* after the Bumpus's hounds devour the Christmas turkey), I said, "Let's go to Paris and have a great meal."

I spied in the Gault & Millau restaurant guide a restaurant in Paris called Pile ou Face (Heads or Tails), with multiple toques, or chef hats, plus all the special wine codes saying they had a great list. This was a classic French restaurant that had wonderful ingredients to show off their cream and butter. Now olive oil is good and good for you, but when you are leaving France and everyone is a bit sad, only real cream, butter, and cheese from Normandy will do. Our last family meal in France was a Poulard de Bresse, a larger Poulet de Bresse chicken in a cream sauce, a filet of beef in a dark reduced cream sauce, and a sole floating in butter that to this day makes my mouth water.

* For fun I looked up Giverny's current hours and found that mountains *can* be moved. *Ouvert tous les jours du 24 mars au 1er novembre 2017, de 9h30 à 18h00. Dernière entrée à 17h30.* Open everyday from March 24 until November 1 from 9:30 until 6:00 pm. Last entrance at 5:30 pm.

Chapter 9

Le Démarrage (Startup)

In hindsight, I can say it is easier and cheaper to start a business in France than many places. But in 1997 after four years in France, I saw the challenges of beginning a business through American eyes. I was to learn, as most entrepreneurs do, starting a business abroad is not for the faint of heart.

France, as an anthropologist would assess, is a culture that is suspicious of business and market economies, therefore of money and success. And money is the American way to keep score, guilt free. Cultural origins of getting taken and a strong socialistic, as in communal or community streak, permeate France. From the regime of the Bourbons to today's government, it is statist in nature, a top-down, Paris-managed economy that wants to control markets for the good of the whole community. The entrepreneur that does not recognize this fundamental difference between unfettered, free, or liberal economies that reward and admire success and the creation of businesses and jobs versus those of a conservative state-managed economy that is genetically suspicious of these same things, is in trouble. Ignore it at your peril.

I often tell my French friends the story of the typical emigrant to America who starts with a food cart, moves to a food truck, and after ten years has a series of thriving restaurants and can now afford to buy a BMW. Americans applaud and say good for you, you deserve it, you have earned it. The immediate French reaction is one of jealously and suspicion that he has been cheating, not declaring income, and does not deserve his success as represented by the BMW. This is why many successful people in France do not flaunt their success and, for example, only drive their nice cars at night or on the weekends away from prying eyes. The fear of being denounced by your neighbors that originated in the Revolution (think of the Guillotine) is alive and well today. God help you if you have a disgruntled employee that you somehow have managed to dismiss. Make sure all your accounts are in order, you have declared everything on your income taxes, and all VAT taxes have been paid, because a knock on the door from the FISC (the French IRS), is not unlikely.

As I was finishing my final exams and preparing for our July move, I began the process of starting my wine business. I contacted a friend Marie-Noel Roux ("Merry Christmas," as her husband teases her). Marie-Noel is perhaps the quietest, most patient, and kindest person I know. She also has a very sharp brain and she is a lawyer to boot. As she led me through various details on the Code Civil de Français, I had a foreshadowing moment thinking there is a lot of stuff in those red books on just opening a business. I wonder what else is in there? Even after four years working and going to school in France being told, "Ce n'est pas Anglo Saxon" (It's not Anglo Saxon), I still was thinking as an Anglo and in common law terms. Simply put, the Napoleonic Code, which is the basis of the civil law code, is based on the premise of a clear codification versus a common law approach of precedent.

In theory, it was not hard to start a business in France in 1997. Anyone could do it, but you needed $10,000 capital and a place of business. Now ten grand does not seem like a lot of money, but in the US at that time any kid with any idea and a $100 (if that) could start a business. No way in France. You had to show that you could actually perform financially. Hence, until recently, there were not a lot of new enterprises and job creation was minuscule. Since 1997, the open borders of the EU have made capital and labor flows easier, but obstacles remain.

The actual incorporation papers were relatively simple thanks to Nancy's Irish passport. Because she qualified as a European citizen, that allowed us to create a business with her as the first gérant (director general). In some ways, the French code civil makes life easy because of its clarity, but we faced one small problem: a place of business. Yes, the chicken and egg conundrum. We needed a place to start a business but how do you get a place of business without a business? No problem. How about a PO Box? Or I can use a friend's house as the address? No way, buddy. You might be a front for someone else. It has to be your place of residence and, if so, we will need a letter from the owner saying it is ok to use their home as your siege (seat or business headquarters). You will need to have a lease showing you exist as a business. Can that lease be for a room? This must be checked, and it must have certain minimum services. Then the lessee will have to declare income. They have to set up a business in order to rent to you. If your head is about to explode or you are muttering something from Kafka, you get it. This is why it was so difficult and still is difficult to get things moving in France. Throw in the popular pas possible (not possible) phrase and it is a miracle I did not pack up before I even got started.

Despite my industry experiences chez Becky, I was undeterred, albeit naïve. I, too, believed that if you built a better mousetrap, the world would beat a path to your door. But I missed the class about how long it would take, and the obstacles I would have to overcome before I began. My business model was to make 5,000 high-quality cases of wine per year and all I needed to do was find 5,000 people in the world who would buy a case or drink twelve bottles a year. It was in selling it not making it that would create a business. Reducing the business model to its lowest common denominator seemed to make for a reasonable theory. But to paraphrase the late Al Hotchkin of the Burgundy Wine Company in New York: making good wine is not the hard part—it is selling the wine. He was the man who placed the first order of my wine. I heard his mantra, understood him, but in no way anticipated the difficulty and the time it would take to become established. When you hear that the wine business is a generational business, especially when you are not trying to make volume, you are faced with an oft-quoted refrain that is all too true: "You make wine for the classes, you drink with the masses."

It is very important to realize that a specialty product, such as Burgundy, is difficult if not a real pain to sell. Burgundian vignerons have a loyal, fanatic, but thin group of buyers. Many of them are old, have full cellars, and thus Burgundy needs to bring in new followers every year. This is time consuming and expensive. Our work as evangelists is hampered by small quantities of wines that are impossible to pronounce and vintages that make the wines taste different every year. We are prohibited from putting the grape variety (Chardonnay or Pinot Noir) on the bottle, except for the most basic wine, while our New World competitors market their wines by grape variety. This can confuse consumers. Finally, do not underestimate how most wines are made to a taste profile. Yes, they are made with grapes (sometimes not), but they are crafted for a specific consumer in mind and to taste the same each and every year. Can we say alcoholic soda pop? Burgundy wine has a subtlety to it that makes each vintage unique.

These hurdles are what makes Burgundy attractive to a narrow segment of disciples but requires knowledge and passion in order to sell. The chain of communication from Burgundy to the end consumer is a long one and requires a great more effort as compared to selling a pallet of fifty cases of chardonnay from California or Australia with a cute Koala on the label that a store can go through every two weeks. A sales rep might have a certain natural attraction to selling the latter than the former.

My plan, based on my experience as a realestate developer, was simple: what is my risk exposure and what can I afford to lose? Thus, worry about the downside and let the upside take care of itself. My plan, based on many spreadsheets and cost assumptions that were soon false, had ten basic goals:

1. Find a building with a long-term lease and get incorporated.
2. Develop a network of grape suppliers and find 60 barrels of wine from the 1996 vintage (wine already fermented in barrel) that I could bottle in the spring of 1998 and immediately sell to get some cash flow.

3. Concurrently find grapes or must (just pressed juice) from the coming vintage, 1997, to vinify and bottle for sale in 1999, 60 to 90 barrels.
4. Buy or borrow barrels and equipment for fermentation and barrel aging.
5. Keep costs to an absolute minimum, live with friends, or barter for a bed.
6. Find a part-time person to help with the barrel aging after the first wine arrives in the fall of 1997 while I am in the States.
7. Limit US costs, develop a US sales network, and raise funds from friends and family for ongoing operations.
8. Keep the total two vintage budget 1997 to 1998 to $200,000.
9. If not successful, pack up and head home and at least say I tried to give it a go. No other American had attempted to make wine in Burgundy.
10. Leave with some great stories to tell and plenty of wine to drink.

My plan, however successful or foolish, meant nothing if I had no address, which meant no papers. No papers meant no business.

Enter Paul Garaudet, vigneron from Monthélie. Big Paul makes delicious wines and is a force of nature: loud, opinionated, funny, and always talks with a cigarette in his mouth. He wanted me to succeed. I told him my frustrations and that I needed to find a locale in order to get incorporated.

"I know of a site in Beaune," Paul said. "It is perfect, let me get the owner's number." The next day, he called me and in his gruff smoker's voice, he said, "C'est Paul, appeler M. Bernard Perrot à Dijon." (It's Paul, call Mr. Bernard Perot in Dijon.)

Monsieur Perot, an accountant, landlord, and vineyard owner, met me at the building the next day. I explained my project, told him that I needed a lease in order to start a business and asked if he could help me. He showed me his building that was perfect for making wine: easy access, a good loading dock, access to a massive two-story tall vinification room, plenty of drains, a large functioning elevator, an office with no heat (but who needs heat anyway?) that with some clean up would work, and a large, cold, damp, and cavernous cellar that could hold over a thousand barrels. The building, 4 rue Jacques Vincent, near the train station, had other charming features I would discover as my adventure progressed—electricity that would short out, an elevator that I learned to fix by overriding said electricity, and an outside Turkish toilet that froze in the winter. It became home. But I must tell you, those first two winters with one electric heater under my sawhorse desk, a first-generation Apple laptop, and a telephone modem, I was chilly and alone in the wilderness.

Perot, an elf of a man who skipped as he walked, had a strange high-pitched voice combined with a bit of a stutter and a nervous laugh. He was disconcerting and difficult to understand, but when he went into his rant of the stupidity of the French legal, accounting, and bureaucratic system, I knew I had met the right guy. He said "Les fonctionaires et les règles sont connes, je vous écris un lettre ca va promettant que je sais faire un bail avec nous. Comme ca, vous pouvez faire les necessaires pour votre society." (The bureaucrats and regulations are idiotic. I will write you a letter promising that I will make a lease with you so you can get your business started.) A handshake agreeing to a five-year lease paying $5,532 the first year, $461 per month with modest increases over the next five and I was in business.

Perot, a commissaire priseur, which is to say a very experienced accountant who others would go to for advice and rulings, ended up being my accountant. Early on in the business, I was frustrated by some obscure regulation and he told me not to worry because it was impossible to know all the thousands of pages of rules and regulations that changed monthly. All you could do was have a defensible logic to your accounting decisions because even the FISC (French IRS) does not understand all the laws.

Hustling to get the lease and incorporation papers completed before we left France at the end of July, I handed Perot a deposit check for the first quarter rent, took the key, did one final quick tour of the building and left with a sense of anticipation and dread knowing that a new adventure had started. There was however one small detail question in the business plan that I did not include in my summary.

The first year of business was what we call bricolage, handyman wine-making. After settling in Boston over the summer, I returned to Burgundy in September for the harvest to procure wine. I started by visiting friends asking if they might need some extra cash and would sell me some wine or grapes and most importantly if they knew anyone who might be selling a bit of wine en tonneau (in barrel)? I worked the terrain relentlessly from morning until night spreading the word of my project. Normally I would get one or two names and then go pay a visit de la part de so and so. This was the key: having the personal recommendation with a pre-phone call or even a personal introduction. These were the votes of confidence, one-on-one credit if you will, which allowed me to get established.

I would often drop by around lunch time, or as the day's harvest was finishing to check out the quality, have a coup (drink), often followed by dinner with the harvest team, old bottles of wine, a scary drive home, a deal in hand, and a hangover in the morning.

With hands shaking, I wrote my first check for wine in November 1997 during the Hospices weekend for two barrels of Maranges 1996 (pinot noir). After the barrels of wine were paid for, I was invited in for dinner for the best oeufs en meureutte I have ever had. This traditional dish of poached eggs in a beef Burgundy sauce was the real deal because the eggs were poached in the beef Burgundy sauce and then served over buttery egg noodles. The meal was like a potion because as I enjoyed it, I realized that this was not theoretical anymore. I was in. Having skin in the game is when you have paid for and taken delivery of the product. Now it was my problem and as my buddy Romano told me early on. "Alex, if the wine is good you get the credit, if it is bad you get the blame. Your name is on it. You own it."

Little did I know that the next summer with a strong economy, wine stocks getting short (1997 was not a large vintage), that the interest in the 1998s would create a perfect storm that nearly did and should have put me under. The devilish irony is that I was able to purchase some fantastic wine, from great producers with prestigious appellations. But what did it matter? Not to tell tales out of school, but it is important to put things in perspective. The following great winemakers back in the late 1990s and early 2000s were still selling off grapes and finished wine because they could not sell all the wine they produced in bottles. The Mugneret-Gibourg sisters who took over from their late father in 1988, Etienne Grivot of Domaine Jean Grivot with important holdings throughout Vosne-Romanée, Sylvain Cathiard of Domaine Cathiard, a tiny producer with superb vineyards was transitioning his holdings from his father, and Pascal Lachaux of Domaine Arnoux, son-in-law of the late Robert Arnoux, all sold me some of their 1998 Vosne-Romanée. Today their wines are Holy Grails but at the time they had the same cash flow problems as I did. This was a business transaction but, it came with one of the highest compliments I could receive. "I would rather sell to you because I know you will do the right thing," they told me. I never took that compliment for granted.

In the winter of 1998, I was on the road developing sales contacts, because if all went correctly after the spring bottling, I would have 1,500 cases of wine to flog. To be clear, once the wines have fermented and are safely in barrels, there is really not much to do to make the wine. Sure, we test the wine, make sure the sulfur levels of the wine are correct to preserve the wine's freshness and protect it against nasty bacteria, do far too many tasting in far-too-cold cellars, perhaps some early bottling, but the real work outside the vineyards is hitting the road to get orders. As a friend said to me, "You've got to move the glue!" No sales, no cash flow, and very unhappy bankers pulling lines of credit. Thus, on the road I went selling my dream because I did not even have a bottle in hand to taste. Talk about hutzpah.

The spring and summer produced some sales, but not nearly enough to cover the 1998 purchases, which made my initial year's budget irrelevant. I made calls for a second round of financing, another $100,000 on top of my first round of $270,000, and I thought I could make it through the 1998 vintage. Yes, I have good friends who were actually pretty smart because they figured the more wine that I bought, the less I would keep if I went under and more for them. Around fifteen high school, college, and business friends bought into the business to get wine at the production cost and figured they were at least three to four times ahead as compared to the retail price.

With money in the bank and a train wreck averted, I was oblivious to the twists and turns that lay ahead.

Chapter 10

Déraillé (Derailed)

Intellectually, I understood that the wine business in Burgundy is counter cyclical in nature, but I grossly underestimated the amounts of capital needed for the machine to function. Winemakers hold stocks of older wine anticipating a sale or an opportunity to borrow against them because of a poor harvest: Burgundy's hidden capital or retained earnings. The system is also built on a chain of mutual assurance that accounts for everyone performing their role: growers, winemakers, négociants, banks, and the markets. At times, they can be in sync and the market is stable with reasonable prices and profits being made along all parts of the chain, but it is rare when everyone makes money. A short harvest, an economic crisis, a weak currency, import tariffs: one or all of these can disrupt the chain. The need to be financially solid with built in capital or access to it is a primordial requirement.

To be more specific, let's examine the typical production cycle of a Burgundy domaine of, say, thirty acres (twelve hectares) from the end of one harvest until the next. The actual year's farming costs are 300,000 to 360,000 euros. The value of the grapes produced, assuming no frost, hail, and a normal crop are 500,000 to 700,000 euros. This can be a much bigger range depending on vineyards: remember, it costs the same to farm per acre regardless of the prestige of the vineyard. In other words, it costs the same to produce a kilo of grapes regardless of the end value of the wine (recognizing that yields vary). If you keep all your grapes, you then can turn the wine into a bottle value of 1 to 1.4 million of gross sales. Not bad, you say. But remember, you are not bottling and selling these wines until eighteen-to-twenty-four months later. In the meantime, you have another growing season to finance, back vintages to sell, and capital expenditures to make—all the while keeping an eye on one or two vintages in your cellar. You do not need an MBA to do the math: the capital needed to run a domaine is tremendous.

In July 1998, I heard that some of the large wine houses were writing contracts with set guaranteed prices for the coming harvest. Although new to the game, I thought that this was an insane maneuver as did my more intelligent friends. We braced ourselves for a train wreck. Throughout the summer and into the early fall, contracts were made with higher and higher prices as a short harvest was predicted along with market demand. Despite the small size of the Côte d'Or, its production is intertwined far better with the world's economy than most sophisticated economic models. In the summer of 1998, the financial markets were strong and not only had they recovered from the financial crisis of 1997, but they were roaring along with a new engine: the dot com boom. I clearly remember the first Amazon commercials at Christmas 1997 and into 1998.

With the 1998 harvest arrived my first alcoholic fermentations. Fermentations are a simple thing when they work correctly and are possible because of yeast. Yeasts are wonderful little machines that eat sugar (grape sugar/glucose) and give off two byproducts: carbon dioxide gas and alcohol—it is that simple. You can buy canned yeasts (think Fleischmann's), dump them into the grapes and juice and voilà, fermentation starts. However not all yeasts are created equal. I promised myself not to get technical or geeky in these pages but suffice it to say natural yeasts are more interesting than canned yeasts. Canned yeasts have been modified to act in a very predictable fashion and give you very predictable tastes. Because wild yeasts are wild and you do not know what you are going to get. This is what makes them interesting. They make more elegant wines because, through the magic of nature and biochemistry, they morph each year and thus give you a wine that reflects not only the indigenous profile of a specific vineyard, but also the sound, timbre, or color of the vintage.

My building had been empty for years and was, from a biological standpoint, dead: no flora or fauna. As a result, the Savigny-les-Beaune Pinot Noir grapes I had bought would not start fermenting. I called a friend and great producer Jean-Marc Pavelot and he gave me thirty liters of his Savigny to jumpstart the fermentation and establish native yeasts in the winery. The next morning my Savigny was fermenting away. Because of the fermentation's slow start, along with a Rube Goldberg collection of equipment and no experience, my first effort at making red was modest. The market, however, was hot and I was able to sell the wine in bulk the following spring to another négociant at a profit. He thought it was just splendid. Go figure.

Not to get sidetracked, but perhaps it is useful here to explain a bit of the vocabulary in the wine business in France and how that differs from that in the US. In the US and all over the world, there is little distinction between who actually grows the grapes and who makes the wine. In the US, for example, the amount of wine that is actually produced from estate-grown grapes is quite small and tends to represent wines at the highest level (prices). In France, when you buy any product, and transform it into a finished product or simply resell it, you are by definition a négociant (a wholesaler). Until recently, estate-grown grapes and wine made from those grapes domaines were held in higher esteem than those wines that we made from bought grapes. Thus, by definition, if you grow 1,000 tons of grapes, but buy in one ton and mix that ton of grapes with the other 1,000 tons, you are by definition a négociant and you cannot call the wine domaine made and bottled. Today, this distinction is much less an issue because many of the top domaines lack grapes to meet their demand so they, too, have developed side négociant businesses.

White burgundy wines were the path I took in 1998 as my friend A.J., a fixture in Chassagne-Montrachet, offered to help me find some must (mout pronounced "moo"), or just-pressed chardonnay juice. Vignerons prefer to sell a fixed amount of just-pressed grape juice than to sell their grapes because there is more precision for the seller. More on this later.

He said, I have a friend in Chassagne, he sells in bulk, a great guy, he will like you and you him. Meet me at 6:00 pm at his place for a coup and a discussion.

On arrival, an "I know you" moment occurred when Philippe Duvernay (Domaine Coffinet-Duvernay) and I pointed at each other and we both shouted out simultaneously "I feel good!" In fact, we had met two years before at a party at the wine school where Philippe's brother was a classmate. At the party, we had several, in fact many glasses of Burgundy and together became buddies and business friends singing James Brown's "I Feel Good!" together. The range of grapes I bought from Philippe in those first years was extraordinary with as many as four different Chassagne-Montrachet 1er Crus: Maltroie, Clos St. Jean, Champs-Gains, Morgeots (chardonnays). We met during harvest at the dinner hour to get the timing and logistics straight for the next day's harvest, exchange ideas, bellowing laughs, and enjoying great food and wine.

Fast forward eighteen months and even though what I made in 1998 was good, in fact better than most, it did not matter because we, meaning all of Burgundy, were trying to sell a high-cost, ugly duckling in the second half of 2000 after the dot com bust, creating our own meltdown. The press panned the 1998 vintage, specifically Robert Parker and his Wine Advocate. With the 1999 vintage, a copious and arguably more attractive vintage in the pipeline and the press already touting its virtues, we were all caught short. The only question we asked each other was how will you manage the loss? The big houses could tough it out or dump their stocks at little to no profit on the French or UK supermarkets to create cash, but my three-year-old boutique business had no options.

The wine business in Burgundy is not a business where you can skip buying from your suppliers one year even if you knew what the economy was going to do, otherwise you lose the supplier. The relationship chain flows like an alternating current. Imagine my reality that as the 1998s are in barrel in the fall and winter of 1999/2000, I am buying multiple barrels of wine from the 1999 vintage. It was a Wild West show of selling, trading, and buying wines I made with other négociants. As an example, I pre-bought from a producer in Gevrey-Chambertin ten barrels of Charmes-Chambertin Grand Cru and six barrels of the Gevrey-Chambertin 1er Cru Lavaux St. Jacques. With this and other wines I made, I was able to trade and buy three other Gevrey-Chambertin 1er Crus, three Nuits St. Georges 1er Crus, four Vosne-Romanée 1er Crus, and six Côte de Nuits Grand Crus. The names of the wines are unimportant, but the significance is that today you can hardly find a drop of any one of the above. In total, I made 255 barrels and 55 different wines. What was I thinking? I was not (later I tapered my wines down to about 35). However, I was thinking that the wines were terrific, there was demand, the bank had extended my line of credit, and as I sold the last of the 1997s, the cash flow would be there and growth would continue along with good margins.

Never mind that the wine business is not the insurance business. We have negative float that from time to time can be best described as the sound of a gaping, blood-sucking wound.

Earlier, I explained the capital production cycle. Here is the ultimate downside: sales of 1998 were nonexistent, 1999s final payments were due by the end of May 2000, the 2000 harvest that must be contracted for is three months away, and the French banks were getting a bit squirrely because they wanted to see real profits, of which there were none. To make life and cash flow even more difficult, I paid a 20 percent VAT on all purchases and seven percent on the grapes. This is one of the great tax rackets in the world, and those who tell you a VAT tax is a good thing (not to mention inherently regressive) are lying. Businesses do not have to pay the VAT tax because it is passed along to the end consumer: it is by definition a consumption tax. What happens is that we pay the tax and then get to recuperate it on a quarterly basis. This creates a colossal paper trail to no good end, except that the government gets a four-month float on the backs of businesses. This, of course, is then built into margins and passed on to the end consumer as a hidden tax in higher prices. To make it even more difficult, a new business does not get to recuperate the tax until they have export sales. At one point in the first eighteen months, I had advanced 80,000 euros for free to the French government.

Bear in mind that French accounting and fiscal policies, although similar to those of the States, have this silly idea that the numbers you put on your tax return, give to your bank and to your investors should be the same thing. In fact, there is one bilan (balance sheet or yearly result) and not a separate set of numbers for the tax return, partner's tax accounting, cash flow number, or EBITA, etc. It reached the point that in the fall of 2000, I met with some of my partners in New York to discuss the business' prospects and they said it was time to pull the plug. I refused. There had to be a solution. I could never give up like this.

The key issue was to show French sales so that the banks could have a cover for their argument in continuing to extend my credit. Oh, I think I failed to mention that French fiscal authorities have the right to declare you insolvent if they so deem. So, how do I show sales in France and move my small amount of cash from one pocket to the next all while making the company look as though it has both solid cash flow and growing sales? My solution: engage The Bank of Alex to raise money to feed the beast while selling the 1998s to my holding company in the States, book the sales in France, and move money around as needed. You do what you have to do. In early 2001, we shipped over virtually all of my 1998s that were in France, nearly 1,000 cases, to a warehouse in Lowell, Massachusetts (totally unlicensed). The elevator never lined up with the floor so to move a 2.5-ton pallet on and off was a major physical act. The goal was to sell the wines for cash so I could live to fight another day. At that point, I could proudly say I had become a rum runner.

This maneuver bought me time, but as sales continued to be slow for everyone in Burgundy, the banks in mid-2001 became even more nervous and started to make noise about calling in lines of credit. Apologies to my banker friends but French banks are no different than American banks. The moment you need an umbrella because it is raining, they ask for theirs back. Shortly after arranging the wine in Lowell, I returned to Beaune to be told by my team that the bank wanted their line of credit back and to declare me insolvent and bankrupt. With my insides churning, I maintained my confidence (which felt Academy Award winning), and told my team as I told my investors earlier, this will never happen. I will pay off the line of credit and operate with my own capital. But how?

For a year, I had been raising additional capital (for a rainy day) in the form of "Wine Bonds," eight-year notes whose coupon interest was eight percent but was payable in wine. (In 2000, the prime rate was in the 8.75 percent range.) The idea was attractive to the holders because they received a decent coupon and could use the interest to buy my wine at the export price, which was at least 2.5 to 3 times less than the full retail price of the wine: $1,000 of interest was worth $2,500 to $3,000 worth of wine in the US. I, in turn, had a real cost of funds of about five percent because I was still selling my wine with my normal margins. This really was a win-win as I raised $225,000 to pay off the local bank, get some breathing room, find the time to make sales, and not worry about getting shut down.

I wish the above history was as tidy as it seems, but the process of shipping and booking the sales to the States was the easy part, because we literally could not give the 1998s away and our cash-and-carry sales were a slow drip. From a business standpoint, I did learn a lesson and drastically reduced grape purchases in order to have a realistic balance between sales reality and wine-making dreams. Bit by bit, I started to sell what I made in a timely fashion so that I never had at any given time more than two vintages on hand: over 2 million euros worth of wine. It was not until 2005, eight years after I started, before cash flow, margins, and production were in balance.

In 2003, I shipped 500 cases of the storied 1999s that were slow to sell even with terrific notes in the press. The economy was still in the tank in 2001 because of the dot com bubble when we were selling the '99s. We finally did sell the '99s in 2004 to 2005 at good margins, even with a pallet partially falling off the loading dock and breaking. Of course, they were all bottles of Grand Cru. Friedrich Nietzsche was right when he said "what does not kill me makes me stronger."

But that was not the final insult. After years of making special offers for the '98s, selling and then schlepping five to fifteen case orders up and down the East Coast for cash, we finally decided to write the wine off. With my tail between my legs, I had to pay to have the wine hauled away and destroyed in 2006.

Chapter 11

Quelle Façon de Gagner sa Vie (What a Way to Make a Living)

Talking about the business of wine should be simple enough. After all, isn't it about making something delicious, having someone excited about buying it, and then enthusiastically connecting great people to great bottles and selling them? This is why I got into the business in the first place: the wine is a sumptuous product that attracts curious people, great stories, has a long history, and is rife with scandals. What could be more fun? Yet as I scribble notes and anecdotes, I become more and more lost. In a moment of shear folly ignoring forty years of experience telling me that writing this book is probably not a good idea, I thought, what the hell, type away, tell the real story.

My years selling Burgundy have taught me that the wine business, especially in the US, to be polite, is a very dysfunctional business. In fact, you have to work pretty hard to create a more illogical, less transparent enterprise that really works at keeping the end consumer in the dark, despite their entreaties to the opposite. A product that should be simple enough to sell becomes lost in its own peculiar language, which places it on a pedestal that it really does not need. Why should a product that gives pleasure, has existed for thousands of years, and is made from the common grape be so complicated? Why does the winemaker feel as though he needs psychoanalysis after trying to sell his product? Why with transparent communications and relatively free and open markets, does wine remain shrouded in mystery? Even as I write this, I feel as though I am Marlow heading up the Congo in *The Heart of Darkness*.

A dissertation on the cost of wine in the world's markets can get quite boring, but a review of its complexity will help us with our bearings. First, most countries place a large tax on alcohol and wine in particular. Whether it is a sin tax, luxury tax, protectionism, or simply a way for governments to milk the cow, this adds greatly to the cost of wine. In France, there is the VAT tax of almost 20 percent and in the UK an excise tax of more than two pounds per bottle plus VAT of 20 percent on top of the excise before any shipping, storage, or margin. Paradoxically, because the governments are partnering in the final cost, this actually (especially in France) keeps reseller's and agent's margins down to ten to 15 percent and distribution on a more rational direct basis.

My French sales were in the 25-to-30 percent range, which is above average for Burgundy, especially since I did not sell any wine to the French grocery stores, which sell 80 to 90 percent of the wine in France. The wine selections in French grocery stores are amazing and each September there is the "Foire du Vin" (wine fair) where the chains try to outdo one another with the best selections. Top Bordeaux growths can be found at bargain prices, much to the chagrin of many who bought the wines as futures (in advance). In general, Burgundy is under-represented and if we see someone trying to sell at this time, we immediately think he/she is having problems.

Asia was a strong market for my Burgundy with Korea, Japan, Australia, and New Zealand being my primary exports. China was on the radar for a handful of collectable Burgundies via the Hong Kong market, but it was a Wild West show that was dominated by the London merchants who set up shop there and who maintain their UK allocations by using Hong Kong and the Chinese market as their Asian London. You can see the efficacy of this, because when Brexit hit, the pound dropped, spiking French wine prices while the London merchants functioned as normal.

The UK market is a very strong one for Burgundy and it cannot be said that the British, who really did create the wine trade, do not have hutzpah. A fellow vigneron said that after Brexit, a London importer asked one of his Burgundy suppliers for a "petit jest" (a small effort) to lower prices because of the drop in the pound (this after three years of hail and a fourth of frost). You can imagine the jest the vigneron said he wanted to give his importer.

This leaves the biggest market, North America, where I sold 40 to 45 percent of my production mostly to the US and then Canada. Canada has a few open provinces whose markets are semi-private but its two principal markets, Quebec and Ontario, are state-run monopolies. Because they are the world's largest purchasers, they have barriers to entry that are formidable for most of Burgundy's small producers.

The US is the largest and fastest-growing wine market in the world with perhaps the lowest taxes in the world on beer, wine, and spirits (excluding the time in 2019 when they placed a 25 percent tariff on French and other European wines). It is worth repeating that the US taxes wine and alcohol at a very low rate, but creates higher prices through a de facto consumer tax created by its rigid import and distribution system. What the US got after repealing prohibition on December 5, 1933, with the 21st Amendment, was effectively fifty-one different countries each with their unique form of alcohol regulation and thus distribution.

In most states we have what is called a three-tiered system: importer, distributor, and retailer/restaurant. There are no true national distributors. But, there are two or three giant distributors who act as oligopolies for the major spirits, beer, and wine brands. Even in this case they are licensed independently in each state and technically buy individually from a national importer or a stateside producer. A national importer or US producer of beer, wine, or spirits can sell to any state but only to the distributor directly. Please note there are multiple mind-numbing permutations with a few states acting as monopolies, others with set margins for the trade, some that allow for the direct shipment of wine from a winery, others that allow you to combine the import and distribution functions under one roof and where I got hooked, Washington, DC, which allows retailers to do all three.

Warren Buffet's Berkshire Hathaway owns a distributor in Georgia. His success and investing style are so well-known that it is no surprise that he chose to invest in a three-tiered system. He invented the term "economic moat," which gives a company a clear advantage over others and protects it against incursions from the competition. The smart guy invests in companies with legal barriers to entry.

As a young man, I learned to make convincing arguments with facts that said otherwise, but I do have to defend the American system because it does cost a lot to buy, store, sell, and effectively deliver a box of wine. Add in retailers and restaurants (especially) who are slow or no payers, with each state having its own payment requirements on delivery, sixty days, state posting for late payments or non-payment, is it any wonder efficiencies are few, costs are high, and there is a constant love-hate relationship between all the players?

Every country has its own peculiar barriers to sales of luxury products. I use the term *luxury product* very purposefully because wines in our price range, $25 and up, occupy a very small part of the wine market. Yes, it is the most conspicuous and "sexy" part of the market, but we do not need wine and especially fine wine to live. What we make in Burgundy and in other fine wine growing regions, such as Napa, Bordeaux, Tuscany, Champagne, and Oregon makes our lives richer in all types of ways, but they are not a basic need.

From a purely technical standpoint, the trade in the US puts all wines over $25 retail in the "Super Luxury" category. Now I too found this and the following fact hard to believe, but the average price of a bottle of wine in France in 2019 was 3.00 euros and in the US is was about $6.50. If you wander the aisles of any grocery store or wine superstore, not including Whole Foods or a gourmet store, but say about thirty miles from any major US city, you will see that most wines are in the $7 to $15 range with the jug wines or large-format bottles occupying most of the space. Average down their cost and it is not hard to imagine that the average bottle is $6.50. Remember, this also includes all the barely drinkable wine by the glass that is served in restaurants and bars throughout the US, not in the establishments written about in wine magazines. This makes up almost 90 percent of the wine market and is the reason why the cost can come down to $2 to $3 a bottle.

In 1999, Frank Prial, the longtime wine writer for the *New York Times*, wrote a seminal article titled, "Why That Bottle Cost $50.00 in a Restaurant." His analysis is as accurate today as then:

- Price from winery: $12.50
- Wholesaler marks up the wine 33 percent = $16.50
- Retailer marks up the wholesale price 50 percent = $25.00
- Restaurant marks up the wholesale price three to four times = $50.00 to $66.00

Add in the euro–dollar exchange rate, tariffs, regulatory costs, or the tax and tip that is now added in for even a glass of wine, and in New York it can cost you $18: when the bottle was probably sold by the producer for $5 to $6.

In response to why the trade's prices are so high, my late Irish and longtime Boston lawyer Bill Brown would say in his classic Boston accent "because they can." Bill was a real character, he was in constant motion as he seemed to sprint between multiple stops on Boston's streets saying, "hi, how 'aar ya." Bill understood the alcoholic licensing business as only a street lawyer can, loved good wine and food, and carried in his pockets just enough cynicism to douse you when you dreamed too much. Bill never stopped schooling me with colorful stories about the booze business that are as applicable today as they have been since Prohibition. Some of my distributor friends would beg to differ, but in their hearts, they know that the following bullet points have occurred many times in various forms.

- Most distributors are highly competitive with one another and there is no real love lost between them.
- A distributor will often make a short-term emotional decision to hurt a competitor that might save him a nickel this week but cost him a dime next week.
- Distribution businesses tend to be family businesses in which the patriarch "ran rum," was street smart, and started the business after Prohibition. His son came in and grew the business and today the grandson and often granddaughter come into the business after getting their MBA.
- The adage "Do not pay the producer until you see tears in his eyes" is true.

One Italian producer after a long day pounding the pavement exclaimed to his importer: "Let me get this straight, I grow the grapes, I make the wine, I put it in the bottle, package it, ship it to you, come to New York and sell it for you, and then I have to beg to get paid?"

The importer replied, "Yeah, you are beginning to get it."

From these humble beginnings very often traced to the end of Prohibition, distributors today run large, sophisticated businesses in a highly regulated environment. It is the business's roots, and the chain of distribution from the source to your table of which I speak. Not so much how does it get to your table, but why is the journey to get there so bloody expensive? After I take you through a few of the gatekeepers, you might find it a miracle as I do that any of us ever get our wines sold and served.

With such a diverse landscape, the constraints to sell any wine is formidable and in Burgundy even more so. When you look at the field of battle in business, you generally know who is fighting with you and who you are fighting against. But what if your foxhole mate and those comrades in arms in other foxholes around you have different agendas, agendas that are all about them getting a medal at your expense?

Chapter 12

Tout simplement, je ne comprends pas (I Am Just Not Getting It)

As I continued to pound the pavement to sell my wines, my salesman's thick skin served me well. I had some decent success with placements, but when it appeared that my wines were a natural fit, no sale. C'est la vie of a salesman, whether one is selling brushes, computer programs, cars, or bottles of expensive grape juice. What I learned the hard way was that most buyers were really clueless about what they were buying and that labels do matter. Sorry to take off the gloves but confirmation bias, the phrase that describes the human tendency to interpret information as supporting our opinion even if the information does not confirm or may even contradict our view, is the reality of all products and the wall over which all salespeople must climb. We underestimate uncertainty and this is why we call for experts to prove the certainty that we all crave. This counterintuitive approach is best described in Michael Lewis's book *The Undoing Project* about the research of Amos Tversky and Daniel Kahneman. "We find ourselves unable to predict what will happen with a great deal of confidence . . . It leads us to believe (and crave) that there is a less uncertain world than it actually is," said Tversky in a *Wall Street Journal* article.

This need for confirmation, certitude or proof that something is good, better, or best is true in all professions and no more so in wine where affirmation takes on perverse forms and proportions. When a winemaker tells you in private that he or she has no idea how his wine will develop in five, ten, or twenty years, but wine experts wax on with a certainty bordering hubris, what are we to think? When I open a bottle, not necessarily mine, and I am genuinely surprised at the depth, perfume, and expression even though the "experts" wrote it off years ago, I am not always surprised.

Once at Ma Cuisine, a wonderful wine destination in Beaune, I suggested we try a bottle of 2000 red Burgundy and my dinner guest, an accomplished sommelier, gave me a look as though I had asked her to worship the devil.

When she was about to pull out her cross to stop me from ordering, like a scene from *The Exorcist*, Pierre, the owner, poured a glass and paused, curious to see her reaction. She was shocked to see a deep ruby red liquid full of perfume pour from the bottle with depth and silky tannins that slid over our taste buds. Because she thought this was a one off, I ordered a second 2000 from another producer and she was dumbfounded. "2000 reds are not supposed to be that good, are not supposed to have color, and not supposed to be good at 7 years old," she muttered. I suppose "they," who everyone was listening to and read when the vintage came out were wrong. In fact, we the winemakers will tell you that we were as surprised as anyone at how delicious the 2000 reds became. In all candor, it was a "light" and tough vintage to make, the grapes did not get very ripe but counter intuitively, the wines gained color, depth, and flavor in bottle. If we cannot predict the outcome, who can?

What is this business that makes everyone in it love the product but recognize that the system is askew? Who are these characters that cannot seem to get out of the way? After a life in the business, I am not any closer to the answer, but I can illustrate aspects and players that led to rueful head shaking.

A THROWAWAY WINE

In general, importers/distributors will have one large professional tasting (an oxymoron if there ever was one) per year, inviting all its trade customers to taste the new vintage and have a meet and greet with its producers. Sometimes a winery rep will be present, normally an attractive young person who has memorized the sales shtick and prays that no one asks a question that takes them off message. For smaller operations, having the owner or winemaker there is crucial to communicate the winery's message and passion.

Our Boston importer's tasting was always in September and I was rarely able to attend because of the harvest. When one harvest was late, everyone asked why I was there and where was Dr. FSG (aka "Dr. Feel So Good"). Jim Alpers, an anesthesiologist extraordinaire, filled in for me for years. Jim was a great rep because he was enthusiastic, loved the product, knew my wines, the retailers and restaurateurs, received some bottles for his afternoon work, and did not have to be emotionally and fiscally invested in the people to whom he was selling.

I too had the same virginal attributes in the fall of 1998 as I prepared for my very first trade tasting in Boston. A combination of fear, enthusiasm, and confidence, however misplaced, accompanied my preparations over the summer for the post Labor Day tasting. By the summer of 1998, importation and distribution was in place so I was able to ship over my 1996s that had been bottled in the late spring as well as several early bottlings of 1997s, such as Fleurie (from the Beaujolais and the Gamay grape).

In the 2020s, Cru Beaujolais are the rage, but in 1997 it was one of those "what are you thinking" moves. Not to give you a history of Beaujolais but the Gamay grape that grows on the granite- and iron-ore based soils in the Beaujolais produces age-worthy wines that in the late 1940s and 1950s sold for the same prices as the best wines in the Côte d'Or. The expansion of Beaujolais south of Villefranche to literally the doorstep of Lyon took its traditional high-quality roots centered around its ten villages and created a mediocre red sea that became Beaujolais Nouveau. This is one of the world's great lessons in brand destruction that is only rivaled by Soave in Italy.

A neighbor of mine in Beaune not only owned 3.5 hectares (8.6 acres) of Meursault but also owned one of the largest holdings at the very center of Moulin-à Vent as well as a lovely holding in Fleurie. Her wine, made with indigenous yeasts, was concentrated, pure, with smooth rich tannins but always has the exuberance of the Gamay grape. The wine in short is delicious, fun, and not expensive but more expensive than the wines from Georges Deboeuf, who was virtually the only reference at the time.

As the tasting began, I rolled into my sales rap and a local retailer, always immaculately dressed, who had and still has one of the most interesting and erudite selections of wines in the Boston area, came up to my table. I poured him some Fleurie, he swirled his glass, tasted, spit, and then said loud enough for me and others to hear "not bad for a throwaway wine." I thought "what the fuck did he just say" and gave him the look. He immediately said "Oh sorry, I did not mean for it to come out that way. What I meant to say was that it was a good way to start your tasting with a wine better than expected." At this point, I am thinking this is not good, I am getting bad vibes about what is ahead, as in years. Should I cry, scream, jump across the table and commit homicide, or simply suck it up? Figuring this was a long match, the first of many, I assumed the defensive position and absorbed the first of many more body blows.

My enthusiasm for Beaujolais was not misplaced because twenty years later, I opened a magnum of said Fleurie and served it blind to some very experienced wine folk. The consensus was that the wine was no more than 10 years old and likely a 1er Cru from the Côte de Nuits, probably from Gevrey.

I AM NOT MAKING BURRITOS

Quincy Steele, the son of Steele wine's Jed Steele, is built like a tight end. He is a gentle, thoughtful young man and winemaker who found himself in Burgundy, fell in love, got married, and became a part of our posse. He and

I talked a great deal, not about technique but about what we are trying to accomplish when we make a wine. This philosophical bent leads to technique but technique never leads to beauty and magic in a bottle.

Reality often intrudes because it is a business and we do have to sell the stuff. This leads to all sorts of discussions about marketing budgets, sample allowances, market visits, the dreaded sales rep "work withs," pricing and the phrase all producers fear more than any other: "price point." The conversation that strikes the most dread in any winemaker is "I need the price of your wine to be at x to sell for y. If it is four percent more, we will sell 15 percent less, etc." These are legitimate marketing questions, but horrible conversations suited for small-scale Burgundy. The emotion invested in Burgundy makes it much less prone to these conversations unless we are talking wines by the glass (suffice it to say a restaurant or bar wants to recuperate the full cost of the bottle with the first glass).

One day Quincy was having a conversation with his father about wine and the pressures the distributor was putting on them to hit a certain "price point." Jed was pushing Quincy to meet the price and Quincy said in exasperation "I am not making burritos! I cannot give it away or change the ingredients now to make a price that works for the distributor."

Did you know that there is a burrito index? Charles Hugh Smith created it to reflect the real cost of goods and their increase compared to the official inflation index that grossly understates the real rate. He writes "I can track the real-world inflation of the Burrito Index with great accuracy: the cost of a regular burrito from our local taco truck has gone up from $2.50 in 2001 to $5 in 2010 to $6.50 in 2016 and $7.50 in 2018—that's a 300 percent increase in seventeen years (inflation rose 40 percent during that period).

So the next time you think your wine is too expensive, think about your local burrito truck and some of the math and margins we have explored. I can assure you that my prices did not increase 200 percent since 2001 but the cost of my grapes have in some cases increased 400 percent.

THE PERFECTIONISTS

Perfection is in the eye of the beholder. But for some reason, man (as in human), especially Americans, have a quest to possess perfect wines as judged by the wine press on a 100-point scale as if we are back in high school and getting graded. I have tasted my share of 100-point wines and I really wonder what all the fuss is about.

I have no beef with any journalist, in fact they have almost always been fair and supportive of my wines. I started to receive very complimentary notes and scores from the French, UK, and US press early but, and here is a dirty secret, it really did not make any substantial difference to sales. Yes, it was important to have good scores, but it did not sell the wine. The logic for a product like Burgundy, because it is so specialized, you can almost say obscure, it takes a long time to get noticed and become real. It is a bit as the logic goes for advertisements and the theory that after seeing an ad ten-to-twelve times you begin to notice the product. My friends in the art world say the same thing talking about new artists and their ability to be around long enough so that they have legs and will endure.

This is the same with wine scores and notes: After you see Alex Gambal in the Wine Spectator, Parker, or the Revue du Vin de France twice a year for five-to-six consecutive years, people began to take notice, remember my name, and I began to exist. After this period, the consumer, of course, first notices the odd high score, congratulates you, asks if they can get some of the highly scored wine, which has been sold out for at least a year because of the time delay between the notes and its release. They also fail to notice the minor miracle that for six years all your wines in the various journals have had good scores that were in a tight range considering the variability of Burgundy's weather. I shrugged my shoulders and accepted it so as not to become a bitter, quixotic figure. What I do get a bit fired up about is the idea that perfection exists, that it is attainable, and worse that it can be bestowed on a chosen few.

There are winemakers who make fabulous wines, have the best vineyards, receive great scores but are horrible people. Since I've learned they are egocentric, and unkind, I will not buy or drink their wine. I see and I viscerally experience the person more than I enjoy their wine (I admit that there are probably people who feel the same way about my wines and me: this goes with the turf). Then there are the wines that are very good, not perfect but behind the wine is a magician, a delight, a personality that fills the bottle with light and that makes joy.

My view, perhaps not rational and too insider, is in the strictest sense more logical if you fundamentally use your brain and emotions at the same time. Wine criticism, as is any criticism (be it music, art, literature, cinema, or food) is relative to that of the critic. As long as the buyer understands this fundamental, we are on solid ground, but the moment they fail to use their senses, we are in dangerous territory.

Practically speaking what are the implications of scores? All the trade says that scores are not important. My high school English teacher would write on our papers when we were filling essay pages with bull, "No swatenbottom."

Some years ago at the Pinot Noir Celebration in Oregon, I was sitting with the Burgundy wine journalist Allen Meadows, aka "Burghound." On the dais, there was a discussion among various journalists about scores and how an 89 is a great score and that 90 is really no different. I whispered to Allan that statistically yes there is no difference, it is less than one percent (0.9888 to be exact), but in wine sales it is the difference between getting bought by a customer or not. What are you going to buy, a bottle whose sales flyer has 89 or 90, a B+ or A-? I protest too much you say, but would you not rather come home as a 6th grader with a test or report card that said A- or B+? The difference is not one percent but is defined as your mother asking if she can make your favorite dinner or saying "That's good dear, you will do better next time."

DEADBEATS

The most surprising and amazing part of the wine business is that it is filled with a bunch of deadbeats. In any other business, you would get shut down. You bounce a check, the bank is none too pleased, it costs you and you do it again, you get called in. You do not pay for your stock trades within five days, you are closed out. You have a margin call, you better have the cash. You do not pay your mortgage for a few months, the bank starts foreclosure procedures. You go out to dinner and tell the restaurant owner you will pay him in six months, he calls the cops. In the wine business, you or your distributor call a starred restaurant in New York or Paris and ask for payment, they say you are privileged to have the wine on their list and you can wait. I have more than once gone into a starred restaurant during full service with my hand out asking for payment making sure customers hear that their classy joint is run by a bunch of deadbeats.

One friend who sold his distributor business some time ago continues to delight me with stories of his travails in collecting payments and negotiating prices. He told me about a restaurant that owed him $1,500 in the early years of his business and wouldn't pay. At the time, this was a lot of money for him. After hounding the restaurant, he decided to play nice, "I took the staff for Christmas dinner at the restaurant and as we walked out, I told them to take the check off what they owed me: it was about a wash." The other was when a restaurant would ask for a discount and he would say "OK, are we talking about a palate (50 cases) or a case (12 bottles)?" It was always a case, so he would turn around and ask them "what would you say if I asked for ten percent off my bill?"

Another time while working with Becky Wasserman, we sold wine to a retailer in Washington, DC, who would not pay. The owner was simply stiffing us because he could, so I sent over a friend of mine, a really big guy to collect. Imagine "Luca" from *The Godfather*, yes, he is built like a refrigerator, intimidating to look at but actually a teddy bear.

He arrived and said "I am here to pick up a check for Mr. Gambal." The owner immediately calls me in France accusing me of shaking him down.

I respond: "Did you order the wine?"

"Yes."

"Was the wine shipped correctly and on time?"

"Yes."

"Was the wine in good condition on arrival?"

"Yes."

"When you sent back the signed purchase order, did the PO say payment in full 60 days from departure cellars in France?"

"Yes."

"Then pay me. A shake down is when there is no product involved." I did get my check but, only after I sent my lawyer with Luca, and had him make sure it cleared.

Things changed for the better in the 21st century. Because of industry consolidation and the risk of getting stiffed is now much lower, although when we begin to work with a new importer and the discussion comes to payment, the first question I always get is can I get an extra thirty days? I think, this is not going to end well. The opposite example was my importer Peter in Chicago, an ex-banker and currency trader who walked in one day, said I would love to import your wines and will pay for them now in advance with my credit card. This got my attention and was the start of a great relationship. Pete has a great sense of humor wrapped in a banker's experience of knowing human nature (if people can get away with not paying they will). He had a neighboring state distributor that wanted to sell my wines who was chronically late in paying his bills, wanted extended terms and after the second order and late payment, Peter cut him off with the "nmdb" rule (no more douche bags).

One of the most surprising realizations and lessons for me was how under-capitalized most of those in the business were. The problem for all wine producers, especially specialists like me, is to find an importer/distributor that can both sell and pay for the wine. The small guys have the passion, know the product, can sell the wine but often cannot even afford to pay for a $5,000 to $10,000 order (I think even if you are warm and breathing you can get a credit card with a $5,000 credit limit). The big guys pay their bills on time, but can do so because they sell so much volume. In fact, they are the bank. Throw in the accounting departments that are constantly billing for samples, wine list placements, extending payments, and who knows what else, it is amazing we make any money.

In theory, we have late fees, but good luck trying to collect them. Years ago, I tried to figure out how the Burgundians coped with the sharpies in the accounting departments who worked every angle holding off paying the bills, and I realized that the Burgundians are even better cash managers and add what I call the "accountant's tax" to their bill in the form of higher prices. Those that pay late or have excessive chargebacks might pay five to ten percent more than another importer to compensate for their payment history, what I affectionately call the "A-Hole Tax." I wonder if they teach this in business school cash management classes? They should.

In defense of most distributors, they are the ones most often dealing with multiple deadbeat restaurants and retailers who milk the system as long as they can. This daisy chain of poor payments is endemic to the business and adds costs to one person: the consumer.

PATSIES

One summer evening, we were at a wine celebration in Volnay with some of the best vignerons in France. As is the habit of Burgundians, each person brings a bottle or two and shares it among other tables. As you would expect, a certain "joie de vivre" quickly develops. My friends, those who you would want in a foxhole next to you, started making a tour of the tables, partaking in wonderful old bottles, making new friends, and reconnecting with old friends (they are Burgundy regulars) when they began chatting with a wine journalist.

My friend and business partner Big Al, a raconteur extraordinaire, saddled up next to said journalist, whose inhibitions were also reduced and started a breezy conversation (as all good interrogators do).

"Can I ask you a question? Each year when you give Alex scores, you give him a range of scores with the caveat that his wines are 'reduced'" (a natural phase in the wine's development and hard to taste and judge the final wine). "I have not seen you give Roumier, Rousseau, Dujac, or Lafon a range of scores, do they not suffer from reduction? You taste their wines at the same time of year as you do Alex's."

"Of course, they do," he replied. "But people expect me to give them high scores." (Roumier, Rousseau, Dujac, and Lafon are iconic Burgundy domaines, they are friends of mine, are humble people, and make spectacular wines. They appreciate the fact that people love their wines. But they don't take it for granted.)

Suddenly our inquisitor could have passed a French roadside breathalyzer. Stunned and realizing that Nietzsche was right, that God was dead, he returned to the table and recounted the story as all of us sat their dumbfounded, aghast, and sick to our stomachs. They say in poker, if you've been in the game thirty minutes and you don't know who the patsy is, you're the patsy. Yes, we are the patsies.

CROWD CONTROL

I love doing wine tastings because fundamentally, I love to teach. My first job out of college as an 8th grade English teacher taught me how to control a crowd of kids, but it did not prepare me for adult crowd control. I can assure you keeping fifteen 8th grade boys in line and concentrating on the subject is a great deal easier than trying to keep adult's attention with booze in their hands. At my tastings, I try to get in a few sage comments before the mysteries of Burgundy's chemistry, aka alcohol, take control of the crowd's inner 8th grader.

Knowing your crowd is a learned craft. At one of my first large dinner tastings, a rather fancy affair on Nantucket, there was a group of ladies out for a girl's weekend who were having their own private party in the midst of the other eighty guests. (A cautionary note to the reader: any wine dinner or lecture over thirty requires one to dumb down one's talk to keep the group's attention.) As I worked the crowd, I went over to say hello and in a light-hearted way said "I am glad you all are having a great time. I can see you all were the class clowns in school." Soon thereafter the ringleader came over, gave me a tongue lashing and said in her Long Island accent: "If my husband was here, he would kick your ass for calling us clowns." An apology ensued on my part and a lesson learned. Since then, I assess the crowd at a tasting as one would look for roadside bombs in Iraq.

Another rule I now follow is to never ever ask directly if someone likes your wine. Like lawyers who do not ask generic questions in court unless they are certain of the answer, asking an opinion of wine even in a small group is a set up. As I work the crowd during or after a dinner tasting I say, "I hope you liked my wines and the dinner." Keep moving as would a politician, or let them tell you how much they liked your wines and say thank you. But above all, keep moving, because you are always at risk of getting attacked by the aficionado.

At my first wine dinner, I asked a table of eight if they liked my wines and it did not end well. The host (a know it all) responded, "Do you really want to know what I think?"

Ignoring the alarm bells deep in my cerebral cortex, I said, "Please, tell me." Even the other guests were embarrassed for me as I dissolved under his withering criticism. I would have been happy with a throwaway wine comment, but I suffered as he made himself feel good at my expense.

I mumbled: "Thank you for coming," and slinked away.

With time, experience, and some healthy avoidance, I have learned to absorb harsher criticism. I have made wines that some people love and others that people hate. This is normal and is how it should be. The problem lies within the aficionado who relies on self-aggrandizement. We know who you are and, as my grandmother from the South would say, "He is an ugly man" (and it is not his looks). I searched a long time for the correct feeling to describe how humiliated I was by the guest's comments and came to realize that I felt like I was the hired help, the entertainment. The guest, because he paid, could say what he wanted. This was only the second time in my life that this strong, sinking feeling occurred to the point where I felt eviscerated. The first was in high school when a coach said, "Gambal, you are never going to make it." The guest, like the coach, only made me strengthen my resolve to never give up, work harder, and prove him wrong.

RULE #6

I have a friend, a real gentleman in the old sense, who has seen all too much of the wine business and says with a great deal of pride and wisdom: "I worked in the business for a couple of years and decided it was not for me." I'll call him Mr. B to protect his identity. Mr. B was part of the crowd of ex-pats from the States and England that descended on Paris in the late 1960s and found a wine wasteland. Out of this vacuum arose the late Steven Spurrier with whom Mr. B was close friends. For those of you unfamiliar with Steven Spurrier, he was the young Englishman who arranged the tasting in Paris in 1976 that put Californian wines on the map. Steven, who I counted as a friend, was nothing like the character played by the late Alan Rickman in the 2008 movie *Bottle Shock.* Steven was always self-effacing, humble, and still got as excited about discovering a new inexpensive wine as much as a great old bottle.

This proximity to and yet distance from the business allowed Mr. B to give me advice, perspective, and preserve my sanity in a business where the actors come and go but their roles and actions remain one continuous rerun. They take several forms, but are fundamentally grounded in the role of a gatekeeper to whom the customer, aka the public, looks for acceptance. Luther and the Reformation always came to mind at each one of my tastings as I encouraged tasters to think for themselves and not look for the approval, or more specifically the interpretation of these self-appointed high priests of wine who believe they are the conduit to true knowledge. This is not to say that some of this erudite class are not helpful, but too often they are merely toll booth operators extracting a fee for their blessing.

In all businesses there are those that are consummate professionals, who know their craft, create value, and give value for money, and then there are those who do it for their self-aggrandizement. None more so than these archetypes:

- Wine consultants who tell people what they should buy, fill cellars, appraise wine, find and create a daisy chain of a handful of collectable "must haves."
- Sommeliers whose wine lists are about their egos and not about either giving the customer value, something unique, or God forbid making a correct profit.
- Hipsters, chameleons who can take on several roles but are most comfortable as the all-knowing sommelier who finds the newest bio or natural wines that often are unstable and will never be age worthy. When asked what she/he looks for, they say "anything that is funky," translated as bad smelling wines that resemble and taste like cloudy vinegar that has been run through a blender (and which I would have been flunked out of wine school for making).
- Bloggers (self-described journalists) that rant, rave, and pontificate about how a wine should have been made, or what the winemaker did wrong, or worse what the winemaker did correctly with no understanding of how grapes are grown, how wine is made, and even worse its context.
- Abused punters, described best by an English business partner who once told me with disgust, that the UK trade takes them out to lunch, tells them how lucky they are to be able to have an allocation of this or that wine, and then five years later the punter looks at his cellar and realizes he likes nothing he was sold.
- Wine merchants and restaurants whose markups are perhaps justifiable on a pure capitalistic basis, but as Frederic Mugnier from Chambolle-Musigny who makes some of the greatest wines in the Côte de Nuits said: "Look, I sell my wine for a lot of money, in fact, more than I ever thought possible. I have no complaints. But when I sell a bottle of Musigny for 350 euros and it costs $2,500 at retail in New York, we the winemakers could never afford to buy that bottle back. Le système est malade et pire c'est cinglé." (The system is sick and worse it's insane.)

In my former and much younger life, I was in the commercial real estate business, which I enjoyed immensely. It was full of characters with massive egos but with risks and rewards commensurate with the business: in other words, there was a lot of money involved. With wine, the money involved is peanuts in comparison, but the egos you encounter make real estate titans Lilliputians. Never have so many become so self-absorbed over something as simple and beautiful as wine.

Early on in writing this book I asked Mr. B to look over this chapter and give me his thoughts (Mr. B was head of PR for Solomon Brothers in its 1980s "Liar's Pokers" days). In his lovely Southern drawl, he said, "Alec ("Alex" in the South), I think you write quite well, and what you write is quite funny. But call me crazy, I do not think it prudent to tell your investors they invested in a shitty business and that your customers are a bunch of assholes. True, but not prudent while you still own the business. Sell the business and then write the book."

Sometime later, Mr. B and I were discussing the above subject and I realized that it is a pity that more people in the wine trade were not raised in Mr. B's family. He told me growing up in South Carolina about his family's rules. Rule #6 was "Never take yourself too seriously." I asked what were the other rules and he said there are no rules 1 through 5 and none after #6.

WORK WITHS

Along with trade tastings and wine dinners, the most dreaded aspect of the business is "work withs." These are customer visits with your sales rep that take many forms—some quite easy and pleasant but most downright horrible. Sales, no matter what the product, can and often do bring out the worst in people. I used to sell parking services, not the most glamorous of products but I developed a thick skin. With wine, the sheer rudeness of buyers of all stripes and the way they treat their sales reps—and reps with their suppliers, who have often come from the other side of the world—is mind boggling. Canceled appointments, no response to calls for an appointment, waiting for as long as an hour when you know the buyer is only trying to bust your chops or is too busy doing inventory (they always seem to be doing inventory). The list of excuses is endless. I like to think there is a special place in Dante's Hell for those wine buyers who have abused their role: I see it as an endless term of doing work withs in a New York summer—climbing and descending subway stairs for long waits or no shows.

I do understand that wine buyers are bombarded with too many requests to taste. There are far too many wines available and any normal person has a limit to sales pitches. However, and you must think that I exaggerate, I can tell you that even the best and most famous wineries often get the same treatment from buyers as I did. I find it amazingly inconsiderate, poor mannered, and just plain rude that the trade acts this way and no one thinks it is unusual. If you do not want to see the rep or the supplier, do not make the appointment and do not waste either their time or yours. Perhaps I am in the minority, naïve, too old school, but reciprocal respect is not a bad thing and when you find a customer of a similar ilk, they become a cherished commodity.

Not to paint everyone in the trade with a wide brush because there are those customers you love to visit, even if they do not buy because they do give you the reciprocal attention you deserve. There was the visit when Ned Benedict, wine director at Charlie Palmer in New York, said that he loved my wines, would love to buy them, but there was no room on his list because he already had the appellations I was selling. His was an honest answer along with thoughtful comments on each wine (later Ned went into the sales side of the business and bemoaned the fact that he now had to sell wine to "assholes like me").

My favorite customers are tasters who quickly go through the wines. They do not waste time because the goal is to taste, learn, decide, and not have the buyer spew his philosophy. The ideal buyer is a rapid-fire taster and spitter who asks a few questions, knows what the store or restaurant needs for the coming selling season, and at the end of the tasting says, I will take x cases of these wines. In other words, a logical, efficient approach that does not disrespect the value of another person's time and ultimately that person.

Then there are those who after going through the whole process say to the rep, "come back next week." Translation: no order. Worse still are those who overthink the wine, who are polite, punctual, you'd like to have them in your home, but sometimes you just want to kill them because they are so slow. This happens often and normally I shrug it off. But one night after grinding it out all day in Boston, I visited my buddy Chris. I was really over it and told him that after ten years in the business, I had had it. I was visiting our "throwaway wine" guy with my newest wines and they were tasting beautifully, real rock stars. One of the problems Burgundy has is "stage fright" and how the wines show especially after a bottling and shipment. Often, they act as in the poem "There was a little girl," by Henry Wadsworth.

Today, however, sex, drugs, and rock and roll were coming forth live from the bottles. I smiled inwardly as our taster and his team took notes. When finished, he looked at me seriously, face perplexed. His comment, amounted to another "are you kidding me" moment.

"I'm just not getting these wines today, I am just not getting it," he said, sending me into apoplexy.

Chapter 13

C'est Son Sac (It's His Bag)

There is probably no other wine or brand that has been so misused and abused than Chablis. It is so different than the Côte d'Or's whites or any other dry white wine for that matter; it remains a mystery. Real Chablis is chardonnay in its most pure, discrete, and nuanced form. I love it. But calling Chablis simply chardonnay is like saying da Vinci painted with a wide brush.

First, a bit of history and a two-and-a-half-hour drive in greater Burgundy, from Chablis to Pouilly-Fuissé. This is a short tour and an interesting story that illustrates the charm, size, and, at times, the maddening complexity of Burgundy.

Chablis (the region), which only makes chardonnay, is 115 miles southeast of Paris and is the first part of the region we call Burgundy. You approach Chablis only after passing miles of gently rolling hills that in May are covered with our sea of yellow colza, rape seed in bloom. If you continue past the Chablis exit on autoroute A-4, for seventy-six miles you mount the back side of the Côte d'Or. You pass Bessy-en-Chaume (Bessy in thatch/straw) at 1,968 feet, the highest point of the Côte and descend directly to Savigny-les Beaune and Beaune, the mid-point of the Côte d'Or. The Jura Mountains are seventy miles in the distance followed further east by Switzerland and the Alps.

If you pass Beaune and stay on A-4 for another nineteen miles south, following the Saône River to your left, you pass to the east of the Côte Chalonnaise, home of Rully, Mercurey, Givry, and its reasonably priced reds and whites. After another twenty miles, a bit after the city of Tournus, you see Le Côte Mâconnais on your right with its delicious and inexpensive white wines from eleven different communes or villages such as Azé, Davayé, Vinzelles, and Chardonnay. Chardonnay is where the greatest chicken or egg debate continues: what came first, the grape or the place? Finally, we reach the back side of St. Veran and Pouilly-Fuissé, whose famous escarpments resemble the American West.

Chardonnay is now practically a generic term for white wine, but I am reminded of the story of when Forest Mars, founder of Mars Candy, was asked if he prayed. He said, "I'm a religious man. I pray for Milky Way. I pray for Snickers." Old timers in the Côte de Beaune and the heart of Meursault, Puligny-Montrachet and Chassagne-Montrachet say a similar prayer. They thank the Good Lord for Pouilly-Fuissé, which made drinking white wine respectable, or let's say habitual for Americans in the mid-1970s. Perhaps in part because it rolled off the tongue easily for English speakers, Pouilly-Fuissé became the hot wine that spilled over into demand for Burgundy's other whites at a time when they could not be given away.

This popularity had its downside for chablis because, as with all things popular, it hit a tipping point that led to over production and some poorly made wines. When I arrived in Burgundy in the early 1990s, the price of Pouilly-Fuissé and Chablis (which has always been a British favorite) were literally the market indices we used to judge the health of the wine market. If there were, and there were at times, literally containers of wine available, we were in the middle of or heading into tough economic times and there were literally no buyers of even quality merchandise (think crude oil in the spring of 2020 when the price went negative). I remember tasting many samples of Chablis and Pouilly-Fuissé in early 1994, and the wines were delicious. But the prices were 20 to 23 francs per bottle including sales commissions for a whole container of 1,200 cases, or around $3.50 to $4.00 per bottle with no takers.

The distance from Chablis to Pouilly-Fuissé is roughly the same as the distance from Washington, DC, to Philadelphia: 140 miles. It is astounding how little wine is made in our seemingly large Burgundian geographical area and how discreet, finite, and delineated growing areas are.

Chablis is tiny, in fact it is one of the smallest fine winegrowing areas in France, yet its size pales in comparison to its history and reputation. According to the Chablis Association there were already 1,200 acres, half the size of Central Park, and 450 growers of Chablis as far back as 1328. This is a much smaller geographic area than the 12,000 acres we now call Chablis, what was generally called the Yonne River growing area. In the early 19th century, the area planted and called Chablis was up to a staggering 40,000 hectares (100,000 acres or approximately 150 square miles), nearly 50 percent bigger than the original District of Columbia and six-and-a-half times the size of Manhattan.

This massive supply, plus its proximity to the Yonne River, provided it with a direct link to the markets of Paris, England, and northern Europe. Suffice it to say, a great deal of this Chablis was rubbish and meant to be drunk and watered down for hydration as a staple on the table. Chablis meant white wine until mildew, odium (powdery mildew), and phylloxera decimated its vines in the 1880s as the railroads opened up other wine regions areas of France to easier and more competitive transportation. By the early 1950s, due to the above, an inability to reinvest, and changing tastes, there were only 1,235 acres planted.

Chablis for all of its earlier successes, could not seem to find a break in the first half of the 20th century, leading to it even now being thought of by the mass market as a cheap table wine. In fact, France negotiated terms at the end of World War I at the Treaty of Versailles to protect its wine areas and prohibited signers from using the words on wines not directly made from the areas. This was the nascence of appellation legislation so that a specific product's name was protected as we would trademark a manufactured product. If you recall from your history class, the US did not ratify the treaty, and this is why we still find "Chablis" and "Hearty Red Burgundy" on supermarket and store shelves. Only in 2006 did the US, with the California's winemakers' agreement, sign a settlement with the French limiting the use of the word Champagne and agreeing to only use Méthode Champenoise on the label for US sparkling wines, but not other regions. Other wines that already existed in the market were grandfathered and thus real Chablis is not the jug wine dreck you find in grocery stores that usually contains no chardonnay and mostly fermented fruit juice.

From the 1950s to the 1970s, times were difficult, but through the tireless work of the cooperative La Chabisienne (probably the best co-op in France) to raise quality along with technological advances, such as better treatments to counter mildew, odium, and sprinkler systems to protect against frost, Chablis inched back.

Chablis' roots, not surprisingly, are in its soil and its fossilized oysters, what we call Kimmeridgean limestone, dating from the Upper Jurassic period over one hundred-and-eighty million years ago. This cool climate, along with its fossilized oyster limestone, creates wines that tend to have less alcohol, and higher acidities that impart a steely and sometimes flint-like quality to the wine. This makes Chablis a wine unique in the world of chardonnay.

Although it is only about sixty miles northwest of Beaune and not far off the Autoroute from Paris to the south of France, Chablis is a world apart from Burgundy. The town of Chablis and its winemakers seem to be part of another era when heartbeats were slower and the seasons were predictable. When I visit Chablis, I know why we came to France. It's France at its roots. There is a sleepy feel that challenges me to look for and find its secrets. Dijon and Beaune present themselves as an open book: the restaurants, hotels, and bars are on every corner. Chablis requires detective work to uncover and get to know its vignerons. The food is hearty but with a certain lightness that reflects the wines, and because it is ever-so-slightly off the beaten path, its food and wines can be real bargains. There are fewer restaurants than in the Côte d'Or and they are spread out. Many sell their wines at small markups.

One restaurant in particular that has one of greatest wine lists in France, is a Gambal state secret.

I wanted to make Chablis but did not for very practical and unromantic reason—I already made enough varieties. I was tempted in 2001 when my California importer, Bob Collins, proposed a Chablis project to me. Bob, with his mad scientist hair, bushy mustache, and experience in tasting and selling Burgundy since the 1970s, talked non-stop. The only way to cut him off was to walk away because you could come back an hour later and he would still be talking.

Bob asked me to visit a vigneron in Chablis who made and bottled his own wine, but also sold grapes, must, and wine in barrels. The domaine of more than fifty acres has some of the most magnificent vineyards in Chablis consisting of four 1er Cru and five Grand Cru parcels. The owner, the late Guy Robin, always wore a cap on his head and had a wonderful Burgundian accent where the letter "r" becomes a word: "airrrrr." My importer wanted to sell some of this domaine's wines but, to be charitable, the quality of the wines was spotty. Bob's project was simple: could I find some barrels of the newest vintage in his cellar for sale that were of decent to good quality, purchase them and take them back to Beaune for a rigorous and precise barrel aging (élevage of about twelve more months), bottle it, and have some good wine to sell?

I called the domaine, outlined the project and one morning in late June 2001, I drove to Chablis with Fabrice, my Burgundy born-and-bred cellar master. We found the domain on a small side street. Even Fabrice was shocked: this was La France Profonde, the old France—the France of his grandparents. Fabrice shook his head and muttered to me, "Ce n'est pas possible, c'est incroyable—quel bordel." (It's not possible. It's incredible—a hell of a mess.)

The winery courtyard was filled with chickens and quite a few big, angry dogs whose poop had made the courtyard even more dangerous to navigate than their snarling teeth. I asked for a brief tour of the facilities. With expectations already low, I was not disappointed as we took in the operation's hygiene and rigor, nonexistent, the bottle storage area, hot, the labeling area, ancient, and the chai, the French word for winery, a dirty barn. We crossed toward Guy Robin's home and my spirits brightened. His garden was full of ripening melons, lettuces, tomatoes, and green beans. At that moment, I knew where Guy's priorities were and wondered, "Can we perform an intervention?"

We descended the steps to his cellar under a mid- to late-1800s French bourgeoisie home—a middle-class farmhouse via some of the steepest stairs I have ever seen. My spirits rose a bit when I felt the air. The cellar was cool and well lit, albeit via a two-line wire method where two live wires were hung in parallel and two metal hooks with attached dangling wires plugged into a socket and incandescent bulb. Today, this method is very chic. You see it in new houses and in New York and Paris boutiques, but the lines there are grounded.

The cellar was stacked two-to-three ancient barrels high, not a problem in itself for Chablis' wines rarely use new wood barrels. They were sealed with wood bungs, barrel stoppers, in the shape of tapered cylinders, cones really or a short belaying pin that you find on the gunnels of an old sailing ship (most of today's bungs, about the size of a silver dollar, are either wood with a piece of inert plastic/paper to complete the seal or of silicone. Both make for a tight seal and can be easily cleaned). As Monsieur Robin had us taste from the barrel, I noticed that there was movement on the bungs. Looking closely, I pointed this out to Fabrice: tiny larvae crawling on the bungs and around the lips of the barrel. Fabrice, a maniac for hygiene and cleanliness in our cellar nearly threw up. To be sure, it is not unusual for stuff (e.g., fruit flies) to get attracted to the bungs when wine goes through its barrel aging, especially in the summer. The wine will expand or contract and thus a slight seepage around the bung does occur attracting some nasty-looking scum, interested flies, and creepy crawlers. If you have ever toured a cave in France, you would understand they are not the most sanitary spaces with mold present on the walls and stone. This is why we disinfect and clean the top of the barrels and bungs with water and liquid sulfur every week to ten days. In this case, there was no seepage and the little worms had just found a nice warm and dark home and did not like being woken to the light.

We continued to taste through his different wines. To be kind, they were boring. I thought we could perhaps clean them up and make them drinkable, but I did not want my name on the finished wine because they could never meet my standards: there were too many mistakes and faults in the wines to make them great. Imagine you have ordered Dover sole or Nantucket scallops from a restaurant and the sole or scallops came out cold, undercooked, or (worse) overcooked. The chef could reheat it or perhaps put more sauce on it, but what should have arrived as splendid dish, representing the quality of the area was ruined because it was not delivered at the perfect moment. Maybe it was because the waiter or line cook was distracted, the owner didn't train them with discipline, or they did not maintain the high standards of the product.

Today it is much less so, but one of the sad facts of the wine business in Burgundy is that in the late 1960s until the late 1980s there was a great deal of rubbish produced. The tide of qualitative improvements in the wine world since the early 1990s has been extraordinary, but there are still many wines that are abject failures and shadows of their vineyard's potential. Growing grapes is hard, terribly challenging; it is a métier or profession all its own, as is making wine, but they are two different functions, not in parallel but connected. Are you a farmer? Do you know how to grow and produce grapes in a climate that is humid, can be too cold or too hot and can be destroyed in five minutes from a July hail storm? Do you know when to pick your grapes at the perfect moment and hand the grapes off to yourself the winemaker or sell them to a winemaker specialist? As a winemaker, do you have the discipline to handle and treat the grapes as a precious commodity, to give them the attention they deserve from the moment they arrive until the perfect moment for bottling? This is exactly the challenge that was presented to us at the end of my year in wine school. It can be seen by twisting the adage: "If you don't break it, you do not have to fix it." Monsieur Robin's wines needed a lot of fixing.

As we were finishing our tasting, I asked Robin "A quelle fréquence faites-vous le remplissage?" (How often do you top off your wines?)

"Chaque semaine" (every week), he replied.

"Monsieur Robin, peux-je redéguster le Chablis 1er Cru Montmains?" (Mr. Robin, can I retaste the Chablis 1er Cru Montmains?) Because the fill of the barrel changes throughout the year due to evaporation, tastings, and tempera- ture, we refill the barrel to reduce oxidation every seven to fourteen days. We also clean the bung and around barrel hole to reduce the risk of acetic bacteria (vinegar) and other nasty things from forming, such as our worm friends. This simple task is literally one of the first things you learn when working in a cellar. Robin's wines were tasting flat and not fresh. I saw that it took longer than it should have for his pipette to fill our glasses and my alarm bells went off. While Monsieur Robin headed to the other end of the cellar to pull the sample, I pulled out bungs from several barrels, and stuck in my index and middle finger. Instead of touching wine, I felt nothing but air.

It is sad but true, I cannot remember how many wines I have tasted - wines that started as magnificent grapes, were fermented correctly, but were ruined by a sloppy élevage. In the fall of 1997 as I was just beginning the business, I tasted a Chambolle-Musigny 1er Cru from a producer I did not know at a wine store in Boston. I thought the juice or grapes were good, but the pro- ducer had botched the wine's finishing (élevage). I contacted my wine broker and asked him, "Does Mr. Chambolle sell wine in bulk?"

He said, "yes."

I said, "let's taste his wines as soon as I am back." At the tasting, I found the young wines, around six to seven months old, terrific, but the wines that were a year older and about to go in bottle were tired and uninteresting. I did the same trick and asked the winemaker if we could taste something older. When he left, my fingers did a quick dive into three or four barrels and as I had done with Monsieur Robin, I found nothing but air. It was clear that, as we say in French, "le matiere primière" (the raw material), in this case, the grapes, was great but the élevage left much to be desired.

The point is that it is difficult to fix sick or broken wines especially when you are dealing with small quantities and that is why we have very skilled oenologists who are at heart doctors of biochemistry and are amazingly talented in repairing wines. Oenologists, our wine doctors, are usually independent laboratories that work for many winemakers. However, larger operations, those making many thousands of barrels, will have a staff of laboratory technicians and oenologists on site whose sole job is to fix and blend wines for their markets. This is vastly different than what I was taught. The goal of the Beaune wine school was not to teach you how to fix wine but rather not to make the mistakes in the first place. This "you don't say" moment is avoided by putting in place good habits so as not to make silly mistakes. It is not glamorous, it is winemaking. It is making sure the barrels, hoses, harvest boxes, press, and floor are clean at the end of each day using lots of filtered non-chlorinated water. Bacteria loves sugar. Thus, if you keep all your operations clean, you reduce to a low level your chance of problems.

We left Chablis speechless about what we found. We asked ourselves: how in the world could someone with such magnificent vineyards turn out such rubbish? More importantly, why and who was buying the stuff? We also made a straight line to the nearest gas station to buy Cokes to wash out our mouths and the memory of our little squiggly friends.

I reported back to my importer that I was sure he could find someone better than me to clean up the wines and that it was not my thing to just make the patient breathe. This however was, and is still a big part of the trade, and not just in Burgundy. The global wine industry to a large measure is a bit like Campbell's Soup; it takes a lot of ingredients to make a consistent and marketable product. You should not be shocked but understand that the industry blends great, good, average, and bad wine into a saleable and often delicious whole. The amount of truly artisanal and small-scale wines that are made is miniscule. Defined by price those wines retailing for over $50 comprise perhaps two to three percent of the total market with those above $100 probably less than one percent. Most wine is made on an industrial or semi-industrial scale. This is not to say it is poorly made or unsanitary. On the contrary, the quality of the wines that are made in this massive scale is really quite remarkable considering winemakers are making a product from a fruit and not adding carbonation to a syrup. They are not competing with the super-premium ($20 to $30), ultra-premium ($30 to $50), luxury ($50 to $100), super-luxury ($100 to $200), and icon ($200 or more).

To his credit, my importer was not to be dissuaded. This is why he and other good importers have a nose for wine. Some of us can smell potential. Just like a good detective, we are both curious and love to look in the back rooms to see what might be hidden.

Bob called, "Alex, can you go back and re-taste? There has to be something there that's good to buy. If not, perhaps you can buy some grapes from him and make the wine for me."

A month later, I headed back to Chablis (Fabrice begged off). This time, I was ready for the dogs and changed my tactics by asking Monsieur Robin if he could show me his vineyards because no one had ever really explained to me Chablis from the ground. We jumped into his 1970s four-wheel-drive Citroën Méhari that today would never pass a safety inspection and bounced through the vines.

Chablis is roughly divided in two by a tributary of the Yonne River with most of the best vineyards, it's 1er Crus on its west bank with a southeast exposures. On the opposite slope are the Grand Crus and some of the best 1er Crus facing southwest, with even better and sunnier exposures: the key to Chablis.

Making great wine hinges on great vineyards—that gives the grower a natural competitive advantage. The various vineyard classifications in both Chablis and the Côte d'Or are not haphazard. In fact, there is a lot of logic behind them.

- The classifications began in the 1100s when the Cistercian monks planted vines and kept them up through trial and error.
- Through trial and error, generations of monks found the best sites for vineyards: sites that drain better, get riper more quickly, resist disease, have less hail and frost, and in the most difficult growing seasons produce the best grapes and best wines.
- The Côte d'Or and Chablis both lie at about the same latitude as Seattle—47.6 degrees north. This is a marginal growing climate because of the hours of sunlight in the summer: gaining daylight hours at an accelerating pace until June 21st and losing daylight hours at a reverse pace after June 21st.
- These vineyards have a strategic advantage with better exposures, SE (many 1er Crus) and SW (many Grand Crus). Their placement captures more of the sun's rays in a growing season than vineyards that are lower or higher on the slope or with less summer expositions.
- Simply put, these vineyards get more even sun that equates to more photosynthesis, and riper grapes so that more of the plant starches (in the grapes) convert to sugar (10th grade biology).

We drove up and down the hills, stopped and walked in his vines and Monsieur Robin pointed out subtle expositions that can only be noticed in situe. "Celui-là est plus frais a cause d'exposition plus est et je le vendanger toujours 5 à 7 jours plus tard." (That vineyard is cooler because its exposition is more east and I always harvest it five to seven days later.) "Ici les vignobles sont tres vieux, l'exposition est un peu moins bien, il faut le garder contre le mildou, mais les vignes donnent des raisins toujours magnifique: concentré et mûrs." (Here the vines are very old, the exposition is a bit less good, you have to be careful and watch for mildew, but the vines always produce magnificent concentrated and ripe grapes.) I entered his lifetime of experience and I inhaled. Most of his vines were magnificent old vines, with well-plowed soils. The life within them was a pleasure to experience. I was in the midst of a vigneron whose passion for this special place was contagious and I, too, caught his virus.

We returned to his house and re-tasted in the cellar. No miracle had occurred since my previous visit, but my detective's nose was keen. I asked, "Avez-vous autres choses à dégustater?" (Do you have anything else to taste?)

He said "Oui, j'ai des choses dans le chai." (I have some stuff in the winery.) Choses means things in French but often for older winemakers whose passion is the vine, the wine is merely an extension so it is not particularly of note. We walked to the hot chai, his winery, aka the barn filled with tractors, piles of firewood, old equipment, and pallets of sugar. The temperature was in the 90s. The dogs followed. As my eyes adjusted, I saw there were five traditional Chablis steel cuves (wine tanks) tucked in corners of the barn between cords of wood.

We moved to the first cuve, the 1er Cru Montee de Tonnere. As he found the key to open the valve, I asked our vigneron when he had last tasted the wine. He thought a minute and said "pas depuis la vendange" (not since the harvest). At this point my spirits fell to new lows. But as he wiped away the dust and cobwebs and opened the value, out poured a pale green/yellow liquid that I smelled, tasted and was transported back to the vines. This was Chablis in its purest and most perfect form. Crystalline, balanced, flinty, with a hint of richness in the mouth and the flavors revealing themselves like the treasure within a Russian nesting doll.

As we moved on to the other four cuves, each one was as distinct but equally different from its kin. This is what we look for in Chablis, but so rarely find. It seemed haphazard, random, and just lucky. The wines were so wonderful. If I had had the money and the distribution, I would have bought all of them in bulk, finished their élevage, and bottled them. When I realized I was in a time warp, I just enjoyed the moment.

After our singular tasting, Monsieur Robin invited me to his house for a glass of 1996 1er Cru that was fresh after having been opened for several days. We continued our conversation at his table covered in laundry with his wife behind the ironing board watching American soap operas (dubbed in French). We talked about what he might have to sell and he told me that the wine in the cuves was already sold. It would be picked up in the fall just before harvest (a typical practice). I asked if he might sell me some grapes from the coming harvest. He said he really did not want to because of a previous bad experience selling grapes and the buyer's haggling over the volumes, price, and quality after delivery (this was no surprise because the buyer was effectively run out of the Côte d'Or for doing the same maneuver to many others). Monsieur Robin said he would sell me must, just pressed grape juice, which normally is a low-risk proposition but looking at his less-than-sanitary conditions, I demurred.

As we continued to talk, Madame Robin mumbled to him, "Je pense que il faut faire le dejeuner pour lui." (I guess we need to make lunch for him.) I freely admit I will never be a translator at the United Nations, but I do understand what is said especially when it comes to food. I was offered a steak—beef or cheval. "J'aime cheval" (I love horse), Monsieur Robin opined. I said beef is fine with me. Madame then began to clear the table and put down paper towels for our place mats and mumbled: "C'est un bordel ici, nous français sont toujours un bordel." (It's a mess here and we French are always a mess.) The French sense of bordel has so many nuances, it always makes me smile. The best way to understand bordel is to imagine if you and your spouse came home unexpectedly one evening and find that your 15-year-old has thrown a party and the entire high school has arrived: this is a bordel.

As the table was set, I asked to use the toilet and am pointed through the kitchen to the petit coin (water closet or WC). Now I hate to say it, but the best way to describe their kitchen is a bordel. There were unwashed pots and pans in the sink and on the roof that adjoined the kitchen sink window along with laundry, washed and unwashed along the stairs down to the equally trashed WC. I gingerly retraced my path to the kitchen and noticed several sacks of fresh (thankfully) meat on the counter and our vigneron poured me another glass.

As Madame Robin walked by with a can in her hand, she asked, "Voulez-vous des petits pois?" (Would you like some peas?) I replied, "volontiers" (willingly) and Madame heated them up in a pot and started cooking the meat in a large cast iron frying pan. As she served us, she asked if I wanted dessert because she had just bought some delicious-looking éclairs and I said "no, thank you." Monsieur Robin leaned over to me and whispered, "Elle est grosse." (She is fat.) At about this time, she reminded the Monsieur that he should not drink too much because of his diabetes and asked him when was the last time he had checked his blood sugar? He waved his hand as husbands of forty plus years do in the "yes, dear" mode. He grabbed his finger pricker, did a quick test as I continued to eat he said, "C'est bon, on boit un autre verre." (It's good, let's drink another glass.)

As we ate our steaks and talked, I faced the kitchen and noticed that the scariest and largest of the dogs—a rather high strung and delightful mix of German Sheppard and Doberman—entered the kitchen, put his paws on the counter, pulled down the bag of meat and started to eat with vigor. In a bit of a stutter I said, "Monsieur, le chien a pris le sac de viande, et l'a mange." (The dog has grabbed the bag of meat and is eating it.) He replied, "C'est rien, c'est son sac." (It's nothing. It's his bag.)

Somehow not surprised, I looked at the 1996 bottle, poured each of us another full glass and washed down the last of my steak with pleasure.

Chapter 14

Courbe d'Apprentissage
(Learning Curve)

There are several questions I am asked over and over again and one declarative sentence that while normal, and logical given the romantic books and movies about making wine, make me go postal.

"Do you make the wine?" I really don't know if I should take offense to this question because it implies that I really do not know what I am doing or that it is part of the next question. Yes, I do make the wine but I did so with a team.

"Do you do this full time or is it a hobby?" Yes, actually more than full time and no, it is not a hobby but a business.

"You are living the dream!"—Seriously?

Since I have at least implied that I made wine for the love of it but also the need to pay the bills, I wish I could rephrase the winemaking question to "Do you cut grapes left handed?" (I do.) "Do you crush them naked with your lover?" (I wish.) "Do you drink coffee and read the *Wall Street Journal* all day?" (This is what my father said he did.)

I dispel all the romantic notions when I talk about how many different directions winemaking pulls me in and for the oddest purposes. Yes, I pick the grapes and wash the barrels out before the harvest, but first I have to drive the truck to multiple vineyards to tend to the vines. This is especially so in Burgundy where vignerons and winemakers with diverse grape sources, must process and vinify multiple tons of grapes from small lots from myriad vineyards. Our vendage, or harvest, took me through a roughly thirty-five-mile area in about six weeks.

My college roommate visited and asked me "How did you learn how to do all this stuff, operate all this equipment, and understand the process?" (This from a guy who developed mobile lab clinics to see inside our hearts.) He was not only questioning how I learned to make wine but also the back of the house processes, the part that no one sees but takes up 90 percent of the work:

bottling lines, inventory control, ordering and maintaining labels, printing back labels for importers all over the world, cleaning and labeling the bottles, preparing them and finally using the forklift, what we call a Fenwick, to load the semis (I have not even begun to mention the tractors, plows, trucks, mowers, pruning, and the diverse equipment that entails managing forty parcels of land over thirty acres). Yep, I did it all, but as the business grew, I tried to hire talent that was better than me.

Little did I know when I started the business in 1997 that the idea of having a small, high-quality winery buying only the best grapes or wine would entail so many diverse functions, not the least of which is being able to fix "stuff." The French word for a handyman or do it yourself is a "bricoleur." Simply put, when you make wine you not only need the equipment, you need to know how to fix it before you call someone who will charge you 50 euros an hour especially when you do not have the money to pay him. On the job training became my new school.

Fortunately, I do like to mess with equipment and fix things, but I draw the line with electricity, where the downside tends to be permanent. With most other tasks, such as plumbing, roofs, masonry, simple carpentry, small barrel repairs and even learning how to override an elevator, it is amazing what you can learn to do with a screwdriver, a hammer and a couple of wrenches, and a ladder. My very first and best equipment purchase was in October 1997—a Craftsman tool set complete with all the screwdrivers, wrenches, and socket sets (metric), which I bought at Sears and schlepped back over to France. The set is still in use daily along with two Mag Lite flashlights used to peer inside barrels.

From the first bottling in 1996 of my Gevrey-Chambertin, which I was petrified I would mess up, I learned about hydraulic pressure. It is incredible how powerful a liquid is even in a small tank under the pressure of gravity and all of the widgets you need to control it: valves, bottling spigots, a stainless-steel float (the same as in a toilet) that controls the aforementioned hydraulic-pressure. You cannot believe how heavy a barrel full of wine is when you have to stack them alone and how stupid you feel when you are racking a barrel with air pressure, forget to release the pressure and you get doused with five-to-ten liters of red wine. It's a mistake I only made once although a variant happened when one of our student trainees was racking a large 450-liter barrel of Bourgogne Chardonnay, did not have the air pressure correctly set, and the end of the barrel literally blew apart. I had never seen this happen before and was stunned by the power of the air pressure. I have broken apart old barrels for firewood and it is no easy task. Luckily, the barrel was just full of chardonnay and not a Grand Cru which normally seems to be the case (insurance covered the cost of the barrel and the contents but not the value of the wine if it were in bottle).

You can learn the basics from a book, but the reality is you have to perform the functions sur place (on site) and make mistakes. This is why at the wine school we spent time working with experienced winemakers who trusted us, but prayed we did not mess up. No matter how much experience or local knowledge you have, there are always lessons for which you are not prepared.

Even artisan winemakers like Etienne De Montille of the eponymous Domaine De Montille of Volnay and Puligny, have learned this the hard way. Once, he told me he bought a parcel of Corton that needed replanting and when he went to see the progress, he realized his team had pulled out the neighbors Grand Cru vines! Aghast, to put it mildly, he found out that the old vines in question belonged to one of his father's clients. His father, the late famed lawyer, vigneron and raconteur Hubert De Montille, then retired, shook his head ruefully with a "You have got to be kidding me look," and said. "Merde, ok, I will fix the problem," he said. (You ought to hear the story about one of their contractor's dump trucks that careened into a neighbor winemaker's garage and crashed through the floor into the cellar.)

I was horrified when much less dramatic mistake was made by my picking team with our first Batard-Montrachet Grand Cru vintage. Here the team picked a row of my neighbor's grapes by mistake. My reaction was the same as Hubert's. I immediately counted the vines that had been inadvertently picked, found my neighbor in Chassagne, confessed our mistake (remember the word to use in French is erreur), and offered a bit more of my vines for him pick than we had picked to make amends.

The questions "How were you accepted" and "Did anyone try to rip you off in a deal?" I answer that I was never refused help by anyone. Almost everyone was positive about my project and proved it by lending and often giving me used barrels and other equipment to help me get started. Of course, there were always a few of my non-core friends and suppliers who tried to get away with what you let them. Often, I would go into a cellar to taste a cuvée and say I am interested in it, how many barrels or surface area of vines do you have available? And what is the price? After the negotiations, I would then say, OK, let's mark the specific barrels and I will sign them and or we mark rows of vines with duct tape so there is no confusion when I come back in three months to pick up the barrel or harvest the rows. Funny how the more detailed the buyer gets, the less detailed the seller wants to be or says "Oh, I meant my second cuvee." Sorry, no sale.

The other trick, and it is one of the principal reasons that courtiers or wine brokers exist, is to make sure you receive what you bought with your contracts. Once, and only once, have I been screwed and it speaks to the hard knocks of the learning curve. In 1999, I contracted to buy from a supplier whom I did trust some Meursault mout or must (just pressed grape juice, a typical Burgundy practice). When the contract arrived, it was not written up with him but with a relative of his who he explained had the same vines as he did. In fact, he farmed the vines and said there was no problem "c'est le meme mout" (it's the same must).

I go for the juice pick up to what is a clean and proper cuverie, but the owners are not in sight. The owner's daughter's boyfriend is in charge. I back up my truck, ask "where is the Meursault?" He points to an unmarked tank, says "la-bas" (over there). "Are you sure?" I asked, getting a this ain't good feeling. Now with hindsight, I realize when he said, "Oui, c'est ca" (Yep, that's it), he did so with the look of a 13-year-old who had just been caught smoking.

My principal cavist Alix de Montille of Volnay, as tough and Burgundian as they come, rolled her eyes when I returned with the juice and told her the short version of the story. The next day after we tasted and analyzed the wine, Alix who in her always diplomatic way said, "les salopes" (those bitches)! The wine was under the minimum sugar/alcohol content for Meursault and it did not come close to the quality of another Meursault we had bought. Born in Volnay next to Meursault, she literally grew up in the neighborhood, knew even better than I did of the reputation of les deux filles (the two girls), and she did what she does best: went on the attack.

She called the girls, told them our concerns, "Peut-être que la cuve était mal étiquetée." (Perhaps the tank was incorrectly labeled.) Alix basically got a stiff arm and some plausible, deniable excuses. Not to be deterred, in good de Montille fashion (her father Hubert and brother Etienne would have been proud of her and were probably coaching her), she said: "On le fait simple, nous devrions goûter notre vin à côté de votre vin et décider s'ils sont iden-tiques ou différents." (Let's make this simple, we should taste our wine next to your wine and decide if they are the same or different.) We arrived a few days later with our sample, descended into their cellars, began to taste their Meursault and it was clear it was not the same. But of course, it was not clear to them. We then asked, can we taste your Bourgogne, a wine a level below the Meursault (so you can disprove our argument).

"Nous n'avons pas de Bourgogne. Nous avons tout vendu." (We do not have any Bourgogne here. We sold it all.) How convenient. Lucky for us, the sisters' patriarch arrived at the Mexican standoff to say hello.

The patriarch, a quintessential Burgundian man in his 80s, energetic, enthusiastic, and full of stories, was a visionary and one of the first vignerons to domaine bottle his wines and not sell to the négociants. He was diminutive with kind eyes and bottom line, a great salesman who, before you knew it, had you buying pallets of wine and reserving pallets for vintages to come. Each of his wines told a story and if you missed it, he would remind you.

He asked, "Quoi de neuf, que faites-vous?" (What's new and what's going on?) What he really wanted, since it was getting close to lunch was a coup, a drink and to flirt with Alix, with whom he had known since she was an infant. Alix, a schmoozer in her own right, looked at me with a sense of hope because different rules exist when the patriarch is present. Alix warms him up, explains the problem with the two wines. He tasted our sample then their sample, reflected for a minute. We say "Les deux vins ne sont pas les memes." (The two wines are not the same.)

"On résolve le problème en goûtons le Bourgogne," he said. (Let's solve the problem by tasting the Bourgogne.) A panic ensued as the girls gave him a hairy eyeballed look and he replied: "Qu'est-ce que j'ai dit?" (What did I say?)

Alix intervened: "Les filles nous ont dit qu'il n'y avait pas de bourgogne dans la cave." (The girls told us that there was no Bourgogne in the cellar.)

He said nothing but gave them a stare that burned right through them, and he left mumbling "réglerez le problème" (work it out).

Now we were in the small problem department. In France and especially in Burgundy you cannot simply change the papers to make things correct because the paper trail is thick with details and suppositions that are based on set yields for each appellation so that the correct taxes are paid and there is no fraud and you get what you paid for. So to régler the problem is not an easy fix, but was settled by selling us the barrel of "Meursault" at the price of Bourgogne and we in turn declassified the wine to the lower appellation so as not to hurt my reputation in its infancy.

Many Americans would say "sue the bitches," but for what, maybe 2,500 euros? How do you prove it? What is the cost? How much upside is there to you and what is the downside? I did not want to appear litigious. What Burgundian would want to sell to a litigious American buyer? The solution was much more elegant and frankly more effective. I simply related in a very low-key way what happened to a few brokers, other négociants, winemakers and friends who in turn, spread the word. Burgundy is small and based on régles or unwritten rules and high degrees of trust that are maintained for the good of the community. This is where I learned that I didn't want to break them at my own peril.

In 2003, the year of torrid heat that no one, and I mean no one in Burgundy had experienced before, different systems were required. We harvested at first light while the grapes were cool and quit around noon before the grapes were over 90 degrees (32 celsius) and our pickers collapsed of heat stroke. To cool the grapes and juice, we used moth-ball-sized dry ice. The nuggets were added to the tanks to prevent the immediate onset of fermentation. The dry ice worked to a point, but the process required constant reapplication and supplies often ran out as the winery resembled an *Addam's Family* set.

Furthermore, if you were buying just pressed chardonnay juice, you the buyer had better make sure the juice was cold and that fermentation had not started before you arrived for the pickup. Why, you ask? Well, once I went to pick up ten barrels of Bourgogne Chardonnay from Sylvain Dussort, a very good source and a buddy in Meursault who had not kept the juice cold enough, or perhaps he was a day late in calling the broker who would call me for the pickup. In any case, the juice was starting to ferment. I was there with my barrels and was in a rush, he was rushed too and we kept asking each other "Ça va aller?" (Is it going to be ok?) "Oui" either he or I replied to each other. Yes, the definition of "une grosse connerie or betise" (stupidity and not engaging one's brain) was about to occur. We both thought fermentation was just beginning and we both cut corners. I thought "my winery is not far from Meursault, there will be a bit of gas in the barrels from the onset of fermentation, but the juice should be fine and there should be no problems."

As we two geniuses pumped the juice from his tank into my barrels, the just fermenting juice and slightly active yeasts became as hopped up as a five-year old returning home with a bag full of Halloween candy. By the time we had filled the first barrel, we had the equivalent of a beer tap that was churning the liquid and delivering foam before it even hit the bottom of the barrel. I did think, "this does not look good" and said, "Sylvain, tu dois ralentir la vitesse de la pompe." (Sylvain, you should slow down the speed of the pump.) After the first barrel was filled and the juice was resembling a nuclear reaction, you would think: "OK you two, this is not a good idea, time to regroup." But, no, we thought it would calm down by the time I got to Beaune and we continued unfazed. In the meantime, those little yeasts were so eager to make alcohol that by the time the last barrel was filled, they were in a frenzy, converting sugar to alcohol and giving off gas. With brains still not engaged, the two of us sealed the barrels with several good whacks of our hammers on the wood bungs so the juice, foam (and gas) would not spill out, I drove to Beaune.

I arrived at the winery, backed up to the loading dock and said to my team "Les fermentations viennent de commencer, cela ne devrait pas être un problème, au lieu de rouler les fûts en place, plaçons-les doucement sur le Fenwick pour le transport." (The fermentation has just begun, it should not be a problem. But instead of rolling the barrels in place, let's place them gently on the Fenwick for transport.) What I didn't realize at the time was there was another problem: we still had to get then out of the truck and onto the forklift by rolling or jiggling them side to side. As we placed the first barrel in the cuverie, I noticed that the ends of the barrels were bulging out a good half-to-three-quarters of an inch, creating a dome like structure. It resembled the first movements of a Jiffy Pop pan as its seams begin to expand and the first few kernels begin to pop. I said, "Merde, regardez le les fonds des tonneaux, c'est comme un tonneau de bière!" (Crap, look at the ends of the barrels, they are like barrels of beer without the reinforced ends!) Yep, I had a problem to the tune of ten highly agitated, in fact, pissed off barrels of chardonnay that were ready to explode.

Fabrice and I discussed what we should do, but we had no time to really reflect because the barrel ends continued to bulge as my *Dumb and Dumber* bung hammering held the gas in place. "Peut-être le gaz va calmer un peu" (Perhaps the gas will calm down a bit), someone suggested only realizing it only calms down when it has some place to go. Great, I thought, we can do nothing but hope the barrels calm down. That was even dumber because the only way for the gas to escape was out the top hole. "Il faut dégager les bonds, il y a aucune autre solution" (We have to loosen the bungs, there is no other solution) as a means to have some control over the mayhem. "Doucement, si'l te plait" (Gently please), I urged the guys as they docilely tapped on each side of the bung like an archeologist uncovering a treasure. It was then that Old Faithful erupted—fermenting juice shot up six to seven feet in the air as we tried to stop the explosion with our hands. It did not subside for a good minute. And this was just the first of ten barrels to unfurl. For numbers two through ten we put buckets over the hole to try to limit the geyser's reach as the bung was knocked loose, but even this was difficult because the force of the gas required two of us to hold it in place as the fermenting juice feverishly escaped. After the first geyser, we were prepared with squeegees and water hoses to direct the escaping juice to the drain. After the cleanup, we figured we lost one barrel (ten percent), much less than we expected—a miracle in itself. When you witness the force and volume of mass escaping from the barrel, a 50-percent loss would have seemed likely, so I considered myself lucky and put this on the top of my list of items of not to do in the future.

Finally, in the spirit of education, I will share another screw up that every reader can benefit from. In 2011, I went to pick up a used Peugeot Sprinter truck that we needed for our new vines with my winemaker Geraldine Godot. Her father, a building contractor, knew the dealer and found us a terrific used one in Châlon Sur Saone, about twenty-five minutes from Beaune, where we could get a much better deal. (For vehicles, contractors, building supplies, you name it. As soon as you leave the Beaune environs and head east or west fifteen minutes, prices drop by 10 to 15 percent before you even begin to negotiate.)

We picked up the truck but the bloody gas tank was virtually empty. (When you buy a vehicle in France, do not expect the tank to be filled.) There was a gas station at the nearby shopping center that you got to by going through an underpass. As we approached, I saw the height sign and said "Es tu bon avec sûre la hauteur?" (Are you sure about the height?)

"C'est bon, ça va, on peut y aller." (It's good, we are clear to go.) As we got closer, I looked up and got more nervous.

"Geraldine, as tu certain?" (Are you sure?) A nano second later, the sound of scraping and compressing metal followed as we became stuck and blocked traffic in both directions. We tried backing up but it only made the situation worse because you back up into the blockage not down from it. As we stood there looking at the truck with our mouths agape, one of the drivers who was blocked by us came up and suggested we simply let air out of the tires. Yep, a memorable moment for sure as we were now freed from our self-induced captivity to drive carefully (doucement), on our the rims for air and gas.

Living in Burgundy may have been "living the dream" to many Americans, but I hope you are beginning to realize it was bit more complex than imagined. It was a rich life filled with some of the best food and wine in the world, but riddled with foreign job hazards. I vividly remember coming back from The Nantucket Wine Festival in May 2002 on the ferry. It was a beautiful New England day with clear air and sunbeams on the water, about as good a New England moment as it gets. After a weekend of pouring wine to folks who would become Gambal wine fans for years to come, I could barely keep my eyes open. Wine tastings are trade shows, pure and simple. We are there to promote our wine and hopefully create new customers: nothing more and nothing less. After about ten hours on my feet for two-to-three days extolling the mysteries and magic of wine, I can get a bit cynical to say the least. There are only so many times you can smile, explain the winemaking process, and say that, yes, we make Chardonnay and Pinot Noir from grapes originating in France, that, yes, they are actually Burgundy grapes, and, no, we do not make Hearty Burgundy or Chablis in gallon jugs (sorry if I am beginning to sound a bit shrill).

The dream is more of a philosophical query posed as a declarative statement. "You are living the dream" could also be posed as "Are you / How do you / How can I live the dream?" The permutations are endless but my answer was as polite as it is now. "It is wonderful to live and work in France to make wine, and I am very fortunate to be making a product I love." This was true—I loved the process, loved the product, loved the place, loved the people, and loved the stimulus I got from running my own small business.

However, in the early years, suffering from classic cash flow issues, such as slow paying distributors, the aforementioned French sales tax advances (VAT 20 percent), home bills, great scores in the press but little demand for my wines, forgive me for telling my closest friends: "Screw the dream, buy the wine!" The point was and continues to be this was never a dalliance and few serious winemakers see it as a dalliance. We love what we do but it is a business and we must make money.

I am reminded of the story in Robert Mondavi's book about his begging for money from his bookkeeper so that he could go on a trip to promote his wine. This was more than ten years after he started his label. My experiences

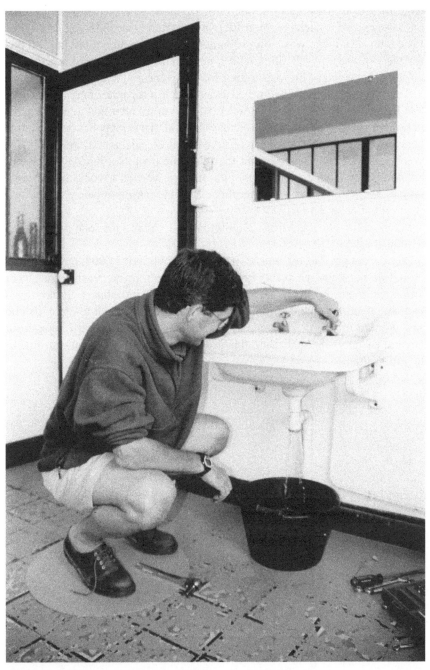

Alex in the early days, acting as a plumber at the first winery.
Courtesy of Carol Alpers

were the same in the 2001 through 2003 period, when I would receive the third degree from Nathalie, my office manager and controller, when I asked for money. My tactic was a bit different because I would not tell her. I'd pay for the expenses myself often with a new credit card and put the expenses through at the end of the year asking for forgiveness and not permission.

To her credit, Nathalie squirreled away cash in a separate company account that I was vaguely aware of but never paid too much attention to until I had a new controller/office manager. One day she said "Tu te rappelles de ce compte bancaire que Nathalie a garde en réserve en cas de paiement d'urgence?" (Do you remember that bank account that Nathalie kept as a reserve in case of an emergency payment?) Nathalie was always worried I would never come up with the money to pay our suppliers. She had reason to worry. I, too, was worried but never let on.

"Oui, celui avec environ 21,000 euros." (Yeah, the one with about 21,000 euros.)

"Oui, mais pas 21,000, mais 210,000." (Yes, but not 21,000, but 210,000.) Now I do not know what this says about me as an owner, cash manager, or an entrepreneur, but thank the Good Lord for having honest employees. I had wondered for eighteen months why I kept sending cash to pay bills and saw nothing in return. The happy ending to this chapter is that I bought our cuverie (winery) in 2003 using this account to fund the renovations, thanks to you, Nathalie!

PART 4

Floraison

Flowering

Flowering occurs in late May or early June and is followed about 105 days later by harvest. It is the confirmation of the fruit's birth. The mornings are warmer, often with breezy afternoons. The fears of a late cold snap and frost are over and we can all exhale a sigh of relief. We might get a cold rain, but the smell of flowers in bloom and the green of the vines, trees, and flowering fruit trees invigorates us and prepares us for the intense vineyard work ahead.

Chapter 15

Le Doigt (The Finger)

What goes on at harvest (vendange), stays at harvest so the saying goes. Harvest exploits are legendary so tracking some of the characters that have come through over the last twenty years helped me track "Gambal Progress," not to be confused in any way with "The Pilgrim's Progress," though there are perhaps some allegorical similarities.

To start, let me dispel a few more notions about the romance of picking grapes. As much as I am asked the living-the-dream question, I am asked "Can I come work harvest and pick grapes?" This question is asked by grown men and women with the dreamy gaze of an 8-year-old holding a new Hot Wheels car or Barbie. My answer is that you really do not want to pick grapes. First, we do not pick grapes, we cut them with small pruning shears that cut novices' fingertips in the first five minutes. To stem the bleeding, you squeeze unripe, acidic, green grape juice and move on. Secondly, our vineyards are not outside a château by a pool. The vineyards in Burgundy are strewn about on hills that can get quite steep, creating amphitheaters offering multiple expositions. They are densely planted on a meter-by-meter basis: basically, over two times as dense as other vineyards. Along with low trellising, the picking occurs in a corridor at two-and-a half-to-three feet above the ground. Once at a wine seminar, a very intelligent man gasped when he saw a picture of our vines of said, "You farm that?"

"Yes sir, that is what we call work and let me explain to you how we are organized and why picking grapes, to borrow a line from Sheryl Crow, 'ain't no disco. It ain't no country club either.'" Peter, a very successful chef and restaurateur in Boston, came and cooked for us one year at harvest and spent a day battling grape vine prickles and hauling boxes. On his return to the kitchen at the end of the day we asked him what he thought of the work and in the classic understatement of a New Englander, he said: "Picking gets the finger."

To understand the work of harvesting, you first must understand some trivia about farming in Burgundy that you can use at your next cocktail party. A

119

hectare is the basic unit of land measurement and is based on two words: hect, as in "hecto," which equals 100 and "ares" which equals 100 square meters. A hectare is 10,000 square meters of 100 meters per side. In Burgundy, we plant our rows one meter apart and the vines within the rows one meter apart, giving us in theory 10,000 vines per hectare. The goal of this dense planting is to produce fewer grapes per vine and to make up the difference by having more plants, forcing competition between the vines, encouraging the roots to go deep both between the rows and between the vines.

In addition, Burgundy has a unit of measurement that few others use in France. It is called an ouvrée, from the word *ouvrier* (worker). It is an old agricultural measurement based on the area a man could work in a day: 428 ares, or 428 square meters. To be precise, there are 23.35 ouvrées per hectare, but we generally just say 24. This is still used as the measurement in vineyard transactions and how we judge yields and production costs. The old term "un pièce per ouvrée" (a barrel's worth of wine per ouvrée), 1/24 of a hectare, was the standard by which the old timers measured their harvest: 24 barrels per hectare or 600 cases of 12 bottles.

Our pruning is primarily a two-cane branch system, called Guyot. One has a longer cane called the "baguette" (as in a French loaf of bread) that will be pruned to six-to-eight eyes (shoots). The second, much shorter cane, the "corson," has two eyes and is trained to become next year's baguette. The vine is pruned low to the ground so any excess vigor goes into root growth for nutrients and the ripening of the grapes.

The vines are pruned and trellised to have a canopy height around 3'11" to 4'3", with the grapes generally located between the middle two of four guide wires at two-and-a-half-to-three feet above the ground. Harvesting grapes is backbreaking work because you are assigned your own row, attack it from the right side, or skiers left. Because most people are right handed, you cut the grapes on both sides of your row by bending down and over the vine in order to reach all the grapes (if you are a lefty like me, you learn to be a con-tortionist). One trick an experienced harvester will employ is to quickly pull off the leaves. This might take a few extra seconds but speeds up locating the bunches and also reduces your self-imposed bloodshed. Can you imagine the carnage if we had harvesters cutting from both sides of the rows? I shudder to think of the workman's comp claims.

The other job that seems easier is that of le porteur (porter) who carries the boxes out of the vines and loads them on the truck. Having done both, I am not sure which is harder. There are several methods used, but we employed a type of wheelbarrow on which we fit three standard harvest boxes. We do not like to overfill them with more that fifty-five pounds so as not to crush the grapes. The grape cases are loaded on the truck and one person can stack the boxes, but with two people the job goes a great deal faster and you get less tennis elbow.

How much can we harvest in a day? In general, and averages are dangerous, we figure about one ouvrée per person per day so we hope to harvest at least one hectare per day with a team of about twenty harvesters and four porters. These rules of thumb are pretty good, but of course heat, rain, and fatigue after a week's worth of work all affect results. Paradoxically, a short harvest can actually slow things down because more time is spent looking for bunches not there. Plus, from a morale standpoint, it is depressing to cover a hectare of terrain and show nothing for your efforts not to mention the chagrin of the owner (me) and my full-time team that have been toiling away for a whole year.

Who do you get? Migrant workers? Nope, it is mostly us pasty European folk doing the work. While African European harvesters are common, migrant farm workers from North Africa are of the distant past. There are companies that subcontract with outfits that bring in teams to harvest with a mix of workers from Eastern Europe but this is for only a handful of the biggest operations. Even medium-sized companies might have a team for the Côte de Nuits and another for the Côte de Beaune each numbering thirty-five which is still not a massive offensive. A typical Burgundy domaine will employ twenty-to-thirty harvesters with another three-to-five as drivers made up of retirees, vacationers, students, seasonal farm workers, and the chronically unemployed along with our apprentices from various wine schools. Finally, we and most others, add workers for the weekend to increase the area covered. However, there is a moment of diminishing returns because you can only process the grapes in the winery for which you have the capacity.

With this as the overview, there is an impression that harvest or crush is a stressful time. Au contraire—I love it. It is the one time a year where there is total focus and we do one thing together: make wine. No tastings, no orders, bills are put on the backburner, accountants don't call, we just focus on one thing: making the best wine we can. To begin, I intensely dislike the word crush because we do not crush our grapes. It is a misnomer. Although there are machines that crush the grapes to release the juice, we do not smash them. Why would you want to crush what you have worked a year to grow?

Now I am probably going to offend someone with the following. It will not be the first or last time, but because we, as in the "Royal We" of Burgundy are small outfits, we do not have fancy tasting rooms. It is we the owners or winemakers who often host. At harvest, because we are to put it mildly, occupied, we say no to visits even though earnest tourists of all stripes think it is the perfect time to visit and taste. If I did get a request, I would say give me a call on my cell phone, given it is still sometimes difficult to call a mobile in France from a foreign phone. If the call came through and I was not up to my ears in grapes or putting out a fire elsewhere, I would say come on by to take a look. Invariably, there is some work to do, perhaps help clean boxes, work the sorting table. Trust me, I will make you useful. Equally, there are moments of sheer boredom when we are waiting for grapes or the press to finish its cycle. If the guest is sympa (nice), a friend or an old client and arrives at lunch, perhaps an invitation to eat with the winemaking team will follow and we will share a bottle usually from an array of countries and producers that have been given to us over the course of the year.

My harvest model began in 1999 and it was not a sophisticated operation. It was an enterprise in its infancy, a mom-and-pop wine show operating on a shoestring. With Fabrice and an intense and at times happily out of his mind Japanese man named Jungi Hashimoto working half time, we harvested my first real parcel of purchased grapes on one-third of a hectare (less than an acre) in Savigny-les-Beaune. The three of us plus my first apprentice, Evan Marley, a few friends, and even a couple of kids (the ultimate way of on doing it on the cheap), managed to harvest the grapes and schlepp the boxes to my 1985 Renault truck that I bought from a classmate for 700 euros and a current inspection sticker. I am convinced it passed inspection because my classmate's brother had his own inspection station. The truck was the definition of a piece of work. The 1985 French electric company EDF, blue J-4, with no power steering, four speed (including reverse) held a lot of grapes and always started. God only knows why we were not pulled over numerous times on our shuttles from the vines to the winery for being overloaded.

Evan Marley, who I had met at a wine and cheese store on Nantucket, worked himself to death. He loved it (I think) and it made him strong as a bull. At about 6:30 pm, at the end of our day harvesting, a very long day, Evan yelled "Alexxxxx."

"Evan, we are about done," I replied. "Look at these amazing grapes, this beautiful evening, and the work you have been a part of." I was being as serious as I could be but knew Evan was over it. "Evan, after this day, operating your own restaurant will seem easy."

The truck limped back to the winery and we finished processing the last of the grapes without Evan. I sent him off to buy a bunch of pizzas, salads, and beer for our starving crowd. As we inhaled our food at the winery Evan said, "You know this is not very good pizza, ours at the wine and cheese shop is much better and I know that I can make an even better one." Evan opened Pi Pizzeria, his wood-fired pizza restaurant on Nantucket seven years later in 2006 and it is some of the best pizza I have ever had.

Exhaustion is a good thing when there are results. I learned a great deal from our 1999 Savigny because it is one of the best wines we ever made. Chief among them, yield does matter. We reduced the yield in July through a green harvest, literally cutting off perfectly good grapes, "dropping fruit" as

Andrew, Cynthia, and James stomping grapes during harvest in 2000.
Courtesy of Carol Alpers

it is called, and we achieved a perfect thirteen-and-a-half sugar/alcohol level, ripe smooth tannins, and a nearly perfect vinification curve. This is the kind of lesson they teach you in school that requires you to do little but to watch the grapes in wonder as they start to ferment after about five days, gently rising in temperature all the while transforming their sugar into alcohol and then gently touching down dry, nature's perfect bell curve: once experienced, always remembered, but rarely duplicated (fast forward twenty-five years, dropping fruit is now rarely done because climatic conditions and vine disease have drastically reduced yields).

I became hooked on grapes and over the next twelve months doubled my efforts in acquiring grape contracts. Because it is easier to handle and make Chardonnay in small batches, the opportunity to purchase Chardonnay grapes was greater. As I expanded my 2000 production, I realized I needed bodies—strong, inexpensive bodies to make this happen. I had the pick of any number of harvest workers because on practically every wine sales trip, I met highly qualified people, sommeliers, bartenders, chefs, retail, and wholesale wine salesmen begging to come over and work the harvest for free. Now this was the period of the enterprise where let's say there was not a whole lot of downside, thus getting busted by the French work police for having undocumented workers was the least of my worries. My one objective was how can

I produce more with less: get bodies, fill them with good food and wine and all will be right with the world.

In 2000, the real harvest circus began with the triple threat of Cynthia, James, and Andrew. Cynthia Meyers, an adorable, smart, non-stop worker bee and sommelier from Nantucket; James Yobski, a serious ex-bankruptcy lawyer who saw the light, quit the law, and was working at The Wine House in Los Angeles; and Andrew Bishop, a bartender at Les Zygomattes in Boston who always had a smile and a joke. They were all unattached.

When the question "where are we going to stay?" came up as it always did, my apartment was the answer. The Beaune Bled (hole in the wall), was the spartan carriage house of my lawyer, Marie-Noel and her husband Remy, who loved having Americans around. If the bled's walls could talk, they would spill a history of housing some of America's better winemakers and France's wine professionals: John Olney of California, Louisa Ponzi of Oregon, Tasha Sprague from LVMH, Soizic Pichon de Sel from Becky Wasserman. There was enough hot water for one good shower, an antique gas heater that I was certain would someday asphyxiate me, a small kitchen with picnic table that was about to collapse, and our "salle polyvalente," (multipurpose room)—living room would be an exaggeration. The carpet was the color of a putting green, the TV had the basic four channels, there was my bed, a pull-out couch, and another space for a sleeping bag. The attic, above the bled, unfinished with no glass in the windows, became the dorm for whoever would brave it. We installed ourselves.

It was a lovely year for chardonnay and with multiple pressing of white grapes, twenty-four-hour decanting, then barrel preparation and fermentation, we would rotate the team in and out of the winery and the fields. Drama we lacked, always a good thing, and I was thrilled I had three serious single adults on my hands who quite quickly picked up the dance of the winery.

Of course, some people understand operations, how things work and their sequence better than others. But with wine, you either get it or not. I have had people who have never been in a winery and in two hours pick up the major process and can anticipate when something needs to be done. I have also had others, very experienced sommeliers, chefs, and even an oil geologist-engineer that never got it or could not physically do the work. This is no fault of their own, but it does indicate that you need a group of people who can work together and who have the ability to choreograph and anticipate each other's steps and roles. In short, a group that knows how to immediately dance well together. These are the people who we say, never need to be told what to do. They are rare gems and I had three good ones on my hands.

A learning moment transpired one morning when Andrew and I loaded the day's pressed grapes in harvest boxes onto the blue truck the night before in order to be disposed of first thing the next morning. As we opened the door, we found the inside of the truck resembling a horror film as fruit flies had hatched, making the windshield and cabin black. We laughed hysterically as we drove sticking our heads in and out of the windows to the distillery to keep from choking on the flies. Our Lucy Ricardo stomping grapes moment (actually a sound procedure) is another favorite memory, along with Cynthia coming out of the shower wrapped in a towel innocently saying "hi," and hanging out to talk to the guys while we were preparing dinner. (We three gentlemen simply shrugged and thought, "God love Cynthia.") Functionality, not modesty, was the word. We all took turns cooking with each person rotating the wine from the cellar to taste blind (guessing the domain and vintage without seeing the label) as well as provide the night's music.

Andrew, who went on to build his own successful import company—Oz Wine Company (with warehouse space in the Lowell that I subleted)—provided the year's musical theme: Disco Inferno. He would crank up the volume and after several glasses of red Burgundy would descend from the bled's attic in his best 1970s disco form. James returned to the law, then again returned to his senses and started his own wine firm, The Old and Rare Wine Company in Santa Monica, California. He became a partner in my vineyards. Cynthia became a banker and has a son who competed as a world cup mogul skier. She came back for the harvest in 2001. While loading grapes in Chambolle-Musigny, it got quite warm and Cynthia being a good, athletic American girl, in a Brandi Chastain moment, pulled off her T-shirt to work in her sports bra. I said, "Cynthia!, and she said, What?" I simply rolled my eyes and carried on. The "boys" from Chambolle were a bit more surprised, actually shocked, and in awe describe it better. To this day, there is always an open seat at lunch in Chambolle for Cynthia.

Chapter 16

Les Règles (Rules)

Around 2007, when my business was on less shaky ground, I began to employ my non-French/EU workers under the auspices of a harvest apprentice sponsored by Beaune's wine school. The school provided worker's comp protection for both the apprentices and the owners. A winery can be a dangerous place with a great deal of machinery that if operated correctly is benign. But mistakes, often due to inexperience and fatigue, make bad things happen. Ventilation is another concern, with carbon dioxide gas being an especially acute danger. Once both the reds and whites are in full fermentation, the cellar can be a literal gas chamber as the heavy carbon dioxide falls to the lowest level. Even though one of my most pleasant moments is when I arrive in the morning and the fermentation smells of yeast and fresh bread, it is also saying danger, we are in full fermentation, be extremely prudent, open all doors and windows and turn on the cellar gas extractor.

Are there any rules for harvest? Only a few, all of which are common sense. But they are strictly enforced. After twenty-two years in the business, harvest became less of a Wild West show because the risks to the business were higher. Not to make a winery sound as though it is a dangerous factory, but a certain cognizance comes with experience and is required to work the harvest. Gambal's rules, which developed over the years and always preceded the harvest, were sometimes affectionately called Gambal's Sermon on the Mount.

Rule 1. What You See, Hear, and Do Here Stays Here

The Vegas adage is not just applicable to the vendange but to all parts of the wine business. Confidentiality is paramount, especially concerning suppliers and pricing. These were the intellectual property of the organization and I protected them fiercely. Making wine, like making sausage, can be messy (you would be surprised what turns up in the tanks after the fermentations). Harvest is a process that is all-consuming, can be physically brutal, and sometimes requires blowing off steam, but always stays in the family (en famille).

Rule 2. Use the Notebook

Each year at the sermon to our apprentices, I would present a small spiral notebook and a pen that they could put in their back pocket. The idea was that as they had questions or observations they could write them down so that we could discuss them together at the end of the day. This could include not only what we ate or tasted, but also how the deliveries arrived and were processed, the order of production, the amount of sulfur used or the way the floor was mopped. The question did not matter. What was important was to take the time to discuss the events of the day together so that we could all learn. This rule proved crucial to the smooth functioning of the operation.

Most normal folks understood the rudimentary concept of the notebook, but one apprentice, an American who never listened to any advice, had little common sense (I still wonder if it was a lack of common sense or simply selective listening). Said apprentice was going to the wine school in Beaune and had worked one harvest in the States (always dangerous). He thought he knew how to do everything better than we did and he told us that daily. Instead of guarding his questions for the end of the day, he posed them immediately as interrogatories. Interruptions like "we did it this way elsewhere," were irritating beyond belief and after several attempts to counsel and warn the miscreant with dismissal, I fired him after six days of work. It had only taken six days before my team would not talk to him, avoided his presence, and told me he was destroying the team's esprit.

Over the years, the little notebooks have become legendary and for those who filled theirs with notes, I gladly provided another one. Matt McClune, who joined harvest in 2001 after working as a cook, bartender, and server at Barbara Lynch's and No. 9 Park in Boston, was a legendary notetaker. Harvest was September that year and so was the bombing of the World Trade Center towers. Matt managed to get over the colossal disruptions of September 11 as did Cynthia, who came back for a second year. She took her place in the attic of the bled, Matt in a corner of the green room, with papa Gambal holding court as we three would open bottles, drink, and reflect on our surreal isolation from the tragedy unfolding in the States. Aware of the world, we turned inward. At night, Matt kept a flashlight on a mattress in a corner of the room writing notes and asking me questions: "Tell me why the two best vineyards of St. Aubin 1er Cru Mergers des Dents de Chien and En Remilly are so different, yet touch one another?" "That bottle of 1987 Vosne-Romanée was still fresh and supple but the 1988 was dry and tired. Why? I thought 1988 was supposed to be a better vintage?" His questions could have gone on all night, but at least they came at the end of the day.

Finally, I would say "Matt, lights out, no more questions. Ask me tomorrow. We have to be at the winery before 7:00 am. Good night." Matt always wanted another story, but I would fall asleep while he wrote his notes and questions under the flashlight's beam. Today his curiosity and precision serves him well as the owner of St. Romain Coffee.

Rule 3. Praise the Chef

Over the years, I had capable people take over more and more of the responsibility of my production and my role became chief monitor. From the modest beginning of figuring out how to afford pizzas each night, we developed a food program that most restaurants would envy. Leading us for five years was Chef Henry from Boston, my friend and dentist who retired early. He became a legend in the kitchen even though the health department should have shut us all down in 2004 for our jack-legged stove that shorted out every five minutes. When it couldn't provide the heat, the meals were shuttled from working kitchens a few blocks away. Along with blind tastings (where we would taste from unmarked bottles and try to guess the vintage) we called "Stump the Chump," early versions of my wines were revisited, not always to great reviews but always to learn. "Risotto Faceoff," Henry's five cheese macaroni with bangers became highly sought after and we all judged inexpensive French supermarket chicken versus its blue-footed noble cousin the Poulet de Bresse in the "Poulet Smackdown." The Poulet contest had a surprising outcome: the locals preferred the expensive Poulet de Bresse and the gringos the inexpensive chicken, but we each mixed them up!

Rule 4. Do Not Be Otis, the Town Drunk

A harvest would not be complete without a night when someone drinks too much and cannot function the next day. This is not to say that there cannot be excess, but there is a fine line between letting it loose on occasion and daily excess where you cannot do your work. To put it in accounting terms, if you drink more than your harvest, you are gone.

Harvest is supposed to be a jolly time and as we became more established, we got many visitors that we would invariably invite for a meal. It allowed me to schmooze (after all visitors are on vacation and they think everyone else should be) without having to devote hours to a client while keeping an eye on operations. Every year there is a guest we name Otis (from the *Andy Griffith Show*) who has been partying all day and wants you to join in. The namesake came from my Southern friend Mr. B., who once deposited his rip-roaring drunk friend Otis on us for a harvest supper, where he continued to pour it down. Mr. B. in an understatement said, "Do not worry, he will not be invited back again."

Perhaps the greatest Otis of all time came when an attractive woman from the Midwest, after several glasses too many, commented in all seriousness on the game of pétanque (bocce ball). "This is a stupid game, why don't we just throw the balls at one another?" This became the quote of the year.

Rule 5. If You Don't Understand, Ask

There was Cara, a ballerina of five harvests who could outwork two men, never complaining while loading and cleaning the press and "Midnight Will," a Kojak look-alike with a mustache, who was a Grand Prix master on the forklift with beer in hand and cigarette in mouth at 10:05 am.

We also had a few characters who just never got it. These included one young man whose family owns Grand Cru vineyards but did not know the vineyard was a Grand Cru and a young woman who said yes to everything (when she actually meant no). When we realized she could not attach a hose to a tank after several years of working at a winery in New Zealand (and her second harvest in Burgundy), we should not have been upset when we noticed one of our tanks leaking wine because she had not closed the door correctly. The French term for clumsy is maladroit, literally bad at getting it straight. This is a much better way to describe these souls.

Rule 6: Show Up on Time or Early

Call me old fashioned, a Neanderthal, guilty as charged, but I believe workers should show up on time. It is not that difficult, requires a small amount of planning, commitment to yourself, and is only respectful to others. Time is a tool with which to establish a working relationship.

With this as context, I did receive workers whom I called projects. These are usually well-educated American 20-somethings, attractive, bright, and polite, but folks who do now know how to hold a broom much less sweep the floor. I know you are thinking that I am exaggerating, but consider for a minute the actual physical process of holding and using a broom. The correct way to use a broom is not to hold the broom in place on the floor and drag it after you but to make small discrete sweeping motions and move yourself once a pile of dirt or debris has been collected. This category of workers included hulking athletic guys who had never weeded or learned how to move a heavy object without causing a hernia but who have a mean golf swing. Yes, I became a cleaning skills coach. I will group them together because there are simply too many good stories to tell. But these workers shared a tendency to blame someone else for their mistakes. This was the case even after many be-on-time lectures.

One night after dinner I told Stephanie "I will be pulling out of the winery's driveway at 7:00 am, not one second later, and I want you on time, which means early." The following morning, she was late and I was livid. I waited until 7:02, drove out the driveway and as I was about to turn the corner, I saw her running after me and the truck. She said she was sorry. "Debbie's alarm did not go off."

With age, enough physical activity, wine, and the occasional sedative, I had learned not to blow my top. I gave her the silent treatment. To the uninitiated that can mean a variety of outcomes: a brief violent explosion, a long and torturous silence that can end in a violent explosion, or a long sermon-like rant that does no one any good. The key is to allow the silence to create fear and attention.

As we rounded Beaune's ring road just in front of the Hotel de la Poste, I said, "Stephanie, let me tell you a story. When I started working for my family's business as a trainee in 1980, I was running garages like all entry-level employees and had a 7:00 am opening shift. The company employed many working-class men and women who were my friends, colleagues, and most importantly my bosses. My boss, Nelda, a 60-something African-American woman with sweeping silver hair, was as sweet as she could be with customers, but a drill sergeant with her staff who she managed in a no-nonsense fashion. The second time I punched in late, she said, "Alex, honey, come here. What's your name? Who owns the company? And who will be here working the early shift long after you have left and are running things in the office? If you want my respect, you have to earn it. To do so, I expect you to be here before I arrive, is this understood?" I replied, "yes ma'am," I did not say another word, and neither did Nelda. And I was never late again.

"Stephanie, who was responsible for getting to work on time, you or Debbie's alarm clock?"

Stephanie looked at me and said, "I understand, I will not be late again." Stephanie went on to get her MBA at Dartmouth and now she's teaching someone else how to proverbially sweep a floor and show up on time.

Rule 7: Get to Know the Mix

Our harvest teams were a mixed grill of French, Europeans, and West Africans. While not the UN, there was a certain self-policing that occurred. Think summer camp, but even a camp director has to be cognizant of the mix. With this warning, here is a brief handbook on harvest hiring, and again, if I offend anyone, please skip ahead.

After our first skeletal harvest teams, we developed a full weekend of harvesting that grew to ten-to-fourteen days. There is an ideal vendangeur who comes for the physical work, to eat, to drink, laugh, and to enjoy the company of others. However, if any one of these activities predominates, or the dreaded shirker disguised in one of the above cloaks shows up, you take a risk of disrupting the harvest team.

Those who come to just work count each extra minute they arrive early or we finish late, and discuss every morning over breakfast (that we have provided) that they need fewer hours and higher pay. They are a royal pain and the problem child you could have avoided had you looked more closely at their work record, noticing that they are either a retired bureaucrat, union employee, political activist (read socialist), but you hired them because you needed a warm body (human resources people, you know who I am talking about).

The eat and drink types we put together because here we have someone who will actually do some good work up to the casse-croûte (literally breaking the crust/bread) or early morning break for coffee, cheese, bread, charcuterie, a cigarette, and wine. Yes, wine does taste great and, no, you are not automatically an alcoholic if you take a couple of nips at 9:30 am. On some of our cold, wet mornings, it does the soul a great deal of good. However, there are always a few who are either alcoholics or they abuse the privilege. When we find that the bottles we pulled out for the harvest are depleting at two times the anticipated rate and the morning's productivity slows to a crawl as others who would not be so inclined to indulge join the 9:30 am festival, it is time to arrêtor le (stop it). The problem is solved by telling the kiddies they have been naughty boys and girls and now the wine is rationed. Self-policing takes over after we tell them if they drink the bottles at one sitting there are no more for the day.

Frivolity characters are strange because they often rest just on the line between useful and annoying. This takes the form of our bavards (chatterers), men and women equally who arrive early, leave late, and never ever stop talking. Always delightful to be around for two minutes, but exhausting after a full day. The key is placing him/her in the rows with complimentary folks.

The chemistry of a team usually has a solid nucleus from the year before with a few new faces recommended by returnees along with a core group of Poles and West Africans. The Poles, our brawlers, who always bring us multiple bottles of vodka are the hardest workers (also the Portuguese) but at times work too fast, leading to resentment as the fastest in the row returns to help the slower non-Poles. The West Africans, primarily from Cameroon and the Ivory Coast show up every day, are delightful, but are talkers and have their own steady if not rapid pace. To complete the collection, we have the shirker who tells you he is the best and most experienced vendangeur. He or she often speaks some English to impress you (brown nose) but you end up firing them because he or she thinks they can run the team better than anyone else and irritates the hell out of everyone.

Like normal human resource issues in any organization, especially when physical work and drinking are involved, Rule 7 took on a special significance that I will elaborate in detail to protect future generations. I will apologize in advance that I love stereotypes because more than nine times out of ten they tell a story better and more succinctly than can a good song, a poem, or a stand-up comic. I am reminded of the Mark Twain quote "Get your facts first, then you can distort them as you please."

Rule 8: Never Hire the Manouches

Harvest 2011 was complicated: I had significantly increased the vineyards I was harvesting, and we had periods of heavy rain that caused us to stop and restart picking. But the bottom line was, I did not hire enough harvesters. In a bid to catch up, I called around to several other vignerons who I knew had finished harvesting and was told by a good friend to call a guy who had a team of North Africans that actually were Romani, technically an Indo Aryan ethnic group commonly known as gypsies or gitans—what the French pejoratively call "Manouches." Desperate for bodies, I called the team's leader.

"Sure, we are available, I have 10 to 12 who will work and we can arrive tomorrow morning."

I said, "Fine, just arrive by 7:00 am so we can get your paperwork done and be sure to bring your IDs and health insurance cards."

The rain fell in sheets all night and I almost canceled picking for the day. But once off and on showers were announced, I stuck with the plan. They arrived at 7:05 and the instant they walked in the door they said, "Avec la pluie, nous voulons plus d'argent." (With the rain, we want more money.)

I said, "Que ce n'était pas une bonne façon de démarrer une relation. Vous n'avez qu'à vous retourner et sortir d'où vous venez." (This is not a good way to start a relationship. You can all just turn around and go out where you came from.)

They immediately backed down and said, "non, non, non, ce n'est pas ce que nous voulions dire." (No, no, no, that is not what we meant.) That then led to three or four of them circling around me as picadors in the ring talking hourly rates, lunch payments, and the like.

Finally, I said, "Vous obtenez le SMIC (salaire minimum) et c'est tout: ça suffit, ou partez maintenant!" (If that's not enough, you can go home now!)

After we disqualified one man in their group who had the same social security number as one man in the others (honestly I cannot make this up), the rest followed our van in their cars. Later, as I walked through the winery, one of our drivers, Rodrigue, a tough, loyal, and good-hearted seasonal employee who is a French version of Joe Pesci looked at me, shook his head woefully and said "Jamais, jamais, jamais embaucher les manouches." (Never, never, never hire the manouches.) I, in my naiveté said, "Ca va aller" (it will be OK). He just turned and walked away shaking his head as I headed out to the vines.

By 10:30 am, the rain coming down, mud everywhere, my normal team willing to continue to work but "Team Romani" striked by not getting out of their cars (I provided rain gear and boots), we called everyone in for the day. Arriving at the winery, I was immediately assaulted by the same three to four ringleaders now asking for cash for their three hours of work. I said, "No, you will get a check tomorrow, not cash, along with pay receipts. If you do not leave now I will call the police." Order restored, our regular crew even more glad to be rid of the miscreants, I walked back to the winery, looked up, and saw Rodrigue standing in the upper winery doorway looking down at me with arms folded, shaking his head and reminding me: "Never, never, never hire les manouches."

Rule 9. Watch Out for Poets

One of the scariest types is not the know-it-all, not the slacker, not the klutz, not even the clueless. No, it is the poet, the philosopher and dreamer who comes for the experience, is capable of doing everything because said poet is intelligent, eager to work, a fast learner, and actually coordinated, but bloody hell thinks things through too much. Because of this and after the 2002 harvest, I put in place a Poet Protocol, where I would sniff out, as the cute beagles do at customs, not food or drugs, but eyes hazy with dreams (rêves). In 2002, my boys arrived: Matt for year two, Midnight Will for the first of five harvests, and a bartender from Boston whom I call "our poet."

Matt and Will understand poetry, but fundamentally they are guy's guys. This was evident the night before they returned to the States from harvest. I, safely tucked into my bed trying to get some sleep, was awakened by the excitement of two, very happy boys retuning in the midnight hours from Au Bout du Monde, the go-to watering hole. Turbocharged energy does not describe it as Will said, "I need some souvenirs for the restaurant where I work in Boston! Let's go get some signs!" (As in road signs.)

Hearing their happy chatter pretending to be as quiet as tipsy boys can be while opening a bottle of wine at 1:00 am, I thought, I have got to get some sleep. "Guys, knock yourself out, go for it."

Matt then asked, "Can we use the tools at the winery?"

I said, "Sure, you know where they are."

"Can we take the Passat?"

"Sure, just be careful driving." I turned and fell asleep.

I am thinking these are two guys who have been terrific workers. What could happen? At the wee hour of 4:00 am, I hear the door open. "Here, give me a hand with this," followed by a "hee hee hee" that only 7 year olds make.

I got up in a fog, opened the door of the green room to the kitchen and I am in Vosne-Romanée. Yes, there is a reason that you can read a highway sign at seventy miles per hour: they are gigantic, especially when it is in a bled. For a nano-second I said nothing. Then: "You did what? In my car? How? Are you all nuts and drunk?"

"It was easy, we had the socket wrenches and stood on the roof of the Passat, it came right down."

"My God, I live here. You could have been seen. I am toast."

"You told us where the tools were."

"I never thought you would be dumb enough to do it."

"Hm, I guess we were. Don't worry, no one saw us. It was 3:30 am." By the way, our poet was passed out in the back seat while the signs were borrowed. I went back to bed, heard another bottle opening and thought, perhaps this is a good time for prayer.

The signs (they had also absconded along with a smaller Beaune sign) made their way back to Boston wrapped in cardboard addressed to No. 9 Park as regular baggage on Matt and Will's flights at no extra charge. Perhaps the most ironic part of the whole escapade is that they were installed in a restaurant in Boston and the Beaune sign was stolen again. The Vosne-Romanée sign is still safely in place. (Just in case, I checked French law to make sure that the statute of limitations has passed to protect my workers and me as an accessory.)

In spite of, or perhaps because Matt and Will were so different, they provided contrast to our poet. Matt, with a year's experience knowing the drill, was managing and explaining basic harvest preparation work to the others. A few days later, all three went with me into the vines for the pre-harvest grape picking and testing (prélèvement), literally grape sampling.

There are two schools of thought on the prélèvement: take eight to ten bunches at random or one hundred fifty to three hundred individual berries from the vineyard. In both cases, the tendency is to pick the best grapes so the rules are to never look at the grapes or bunches and pick high, low, left right, in and outside of the bunches in order to get a valid random sample. The purpose of this is to begin to examine not only the sanitary, sugar, acid, and pH levels of the grapes but also to begin to chart the rate of change of all of the above: in other words, the grape's rate of maturation.

The four of us jumped into the indestructible 1991 VW Passat wagon, and drove south to Chassagne-Montrachet at 8:00 am. With a bit of hustle, we could finish in the early afternoon in the north covering twenty-five or so vineyards. Now this is not hard work. It is actually quite enjoyable to observe the vines, talk about the vine's health, compare them to other vineyards, and begin to get a feel for the year. "You can even lick the rocks and see if there really is a difference in the terroir," says Will. It takes time to cover such a large geographic area, so when I attack the prévèlement and give each guy a plastic bag to fill with fifty-to-seventy-five berries, I want them to move. Most get the hang of this quickly and, at a minimum, when you see me moving along at a rapid clip, and I do like to walk fast, most get the hint and adapt. At the end of the first row of the first vineyard, I looked back and saw that the poet had barely covered a quarter of the row. I muttered "what the hell?" and yelled, "Come on, get moving, we have thirty vineyards to go."

"I know, I know, but I want to savor the experience," said the Poet.

"Savor, schmavor! Hurry up and pick the goddamn grapes and quit slacking and get in the car, we are going!"

I should have known that along with being a poet, he was a romantic one at that. Not in the Walter Scott mode, but more in the line of a French poet such as Rimbaud. I had an American that was intoxicated on harvest. I thought, how I am I going to get this guy to do anything? Scream, cajole, shame, get the guys to bust his chops? Nothing worked and each day he moved more slowly, writing more feverishly in his notebook (my notebook) all the while dreamily pretending to work. A metamorphosis had occurred before my eyes: a seemingly hard-working bartender from Boston in three days had become nul (useless). To make matters worse, he had a crush on one of my new employees. As they say in Texas "useless as tits on a bull."

As our poet tried to get the attention of "little missy," who I am not sure ever really paid him any notice, Matt and Will tried to shame him into pulling his weight. He was oblivious to his surroundings and had created his own harvest movie all the while asking Sam and Will, "Do you think she digs me?"

For our end of harvest celebration, we decided to treat our team to a real American barbeque. Sam, Will, and our Poet, who we had to drag along to at least help carry back the provisions, began by going to Saturday's market and visiting Pascal, the boucher sympa (friendly butcher). He was exactly what you might imagine a French butcher to look like. About 5'10" and stocky, with massive forearms, mutton chop sideburns, a full head of slightly reddish hair, a ruddy face, he was welcoming, always interested in what you were cooking and how you were going to cook it. For years, he had become our go-to source for organic free-range turkeys at Thanksgiving, the most delicious and outrageously expensive turkey in the world. He would special order them for us in early November so we get them at the end of the month since turkeys in France are raised for Christmas. Over the years as more Americans settled in Beaune and celebrated Thanksgiving, he became everyone's supplier.

Our visit to Pascal was difficult as we surveyed the chicken yard birds, poulards, poulets de Bresse, short ribs from free-range pigs, quails, pigeons, and cuts of Charolais beef, the regional white cow of Burgundy. After serious debate with Pascal over the various merits of barbequing or smoking meats, we decided on a half quail and half a pigeon for each person plus slowly smoked brisket, poitrine (breast or lower chest). The boys began smoking the brisket on my Weber kettle grill in the mid afternoon as the last grapes were arriving, driving the team mad with its smells. Finally, roasted potatoes, homemade coleslaw (vinegar based), four different local cheeses, baguettes, and raspberry fruit tarts from our favorite pâtissier completed the meal.

Quantities of my Bourgogne Pinot Noir and Bourgogne Chardonnay began the evening for the team that totaled thirty hungry souls. As the wine continued to flow, I brought up other bottles from previous vintages from the cellar. Announcing the new wine to the gathered was a Burgundy tradition of respect and a great way to thank the team.

The talk became faster and louder as we tried to engage the Poet, but he had become a real third wheel. He gazed dreamily at his beloved the whole evening, becoming more and more sappy as his body was warmed by Burgundy's liquid magic. Toward the end of our fête the poet posed one of his frequently asked existential questions: "Do you think she digs me?" Will, replied in his best Jerry Seinfeld pitch "Yeah, man she reallllyyyy diiiggs yoooo!"

Chapter 17

Deux Vies (Two Lives)

Traveling back and forth across the pond from Boston to Burgundy took its toll on me and my family. Over the years, I have thought a great deal about how could I have built the business without commuting, and in retrospect, if we had all stayed in France, the business and family model would have been totally different. Sales, financing, sourcing grapes, and operating costs would have had a much more European focus, especially to the UK and Paris. US sales would have still been significant but most likely 40 to 50 percent as they became after twenty years, rather than the virtual 100 percent they were in the first five to eight years of the business. My sources of capital to expand would have been limited and this would have had a natural braking effect on the business and especially my grape purchases, leading to a happy circle of actually having the capital I needed for current production. In other words, I would have bought just a bit more than what I had orders for, rather than the riding the rollercoaster of financing that led me to get creative with investors. Finally, if I had been present (sur place), operating costs, employee labor and their significant French social costs—a good forty-five cents on every euro paid—would have made my balance sheet a great deal easier to manage. But rewriting history, although a good exercise, is just that. My anything-is-possible-so-don't-slow-down world view prevailed, while I commuted from the Beaune bled and built the business.

Our choice to relocate to Boston rooted Nancy, Tyler, and Alexa. By 2002, Nancy was flourishing in her job as head of graphic arts and advertising at a local silversmith, Tyler was a sophomore at the University of Southern California interested in language and world music, and Alexa was a junior in high school, thriving in the classroom, socially and with her sports, field hockey and lacrosse. My world headquarters resided in our unfinished basement with Alexa's pet rabbit, Bebe, running free and Tyler's chinchilla, Chini, exercising on his spinning wheel non-stop. My children were allergic to most cats and dogs thus we had a history of rather exotic, hypoallergenic, low-maintenance pets.

I traveled to Burgundy eight to ten times a year for ten- to twelve-day stretches and adhered to this schedule except for harvest when my absence would be for four to six weeks. I normally had a July trip when I would finalize the various grape contracts and put together a general harvest plan with my team and then everyone would go on August vacation. Because les vacances d'été (summer vacation) are so sacrosanct in Europe, and the major work in the vineyards is completed toward the end of July, there was always a solid month or more when business slowed to crawl and I could monitor the season's progression from the States via email.

The family farmhouse on the Cape was a great respite. The kids got summer jobs and Nancy would work in the garden. When I was home, we would go to the beach, then come running back at 4:00 pm when the wind ceased to keep the greenheads at bay. For dinner, we would buy fish and lobster just off the boat at the local fish market and stop at Fancy's Farm stand for Silver Queen sweet corn. More often than not, our family dinners became a time to entertain on the closed in porch so we wouldn't be eaten alive by the mosquitos.

The two lives I was leading were depleting. I loved being with my family in the States, but when in Boston or on the Cape, I missed Burgundy: my business, my team, my friends, and the adventure of creating something that had never been done before. When in Burgundy, I was a lone wolf, I missed my family and began mourning the life we had in France together that was gone as the kids got older. I wanted it back but knew it was never going to return. Our family unit felt more distant with each day. A split personality is too dramatic to say, but I had a split life—one in Burgundy and one in Boston. This wreaked havoc on my marriage to Nancy. She was content in her life. She played tennis with friends and had plenty to do when I was out of town and all seemed well on the surface when I was there. But my absences became longer, my returns more difficult. We were living separate lives—together, but apart, intersecting but interacting less and less. It was clear to me the marriage would not last and I reached a point in the fall where I did not want it to. My passion was elsewhere. I had found it in wine, in Burgundy, and I could not and would not let it go.

For several years, I had given tastings to bike tour groups from Vermont Bicycle Tours as a tool to sell a bit of wine, but most importantly to create long-term customers for my wines. Alisa, the European director and a good friend of Nancy's and mine steered many tours to our door. In the middle of harvest 2002, she asked me to host a tour and tasting for a private bike tour group for the president of the company.

"You know that I never do tastings and tours at harvest," I said. "This is not a good idea. There is no worse moment for me to receive guests."

"Alex, you owe me one, if not more than one. Do it." You can say a great many things about Alisa but she is always clear, direct, and usually gets her way.

The tour's leader was Diana "Dee" Williams, a coach for the US Olympic Freestyle mogul team who led bike trips in the off season when she was not coaching. I honestly did not notice her at the tasting, much to everyone's surprise, because I really did have other priorities: I was in the middle of the harvest, keeping an eye on my team and grapes were arriving all day. However, I did my duty and made time for her, hosting a good tasting and an entertaining session of show and tell. At the end, Diana asked if she could call me and ask me some questions about the industry and the wine marketing program in Dijon. "Sure," I said offhandedly.

About a week later in the middle of pumping over some wine in tank, Will yelled: "Hey Alex, there is some girl on the phone that wants to speak to you. I think it is the blonde from the bike tour last week. Maybe she digs you!"

"Ha ha," I replied, "You bastard. Oh God, just what I need, another chick that wants to get in the business." I wrote down her number, the date and time we were to meet next to the phone on the unfinished plasterboard loading dock and did not give it much more thought.

Diana arrived on a Sunday, professionally dressed, notebook in hand, brimming with questions, I in my shorts, purple-stained T-shirt, Teva sandals, and fermenting wine on my hands. I greeted her with a cup of coffee with cream (cream in coffee in France is not done, but as I came to learn Diana loved cream in her coffee). She explained the marketing program to me, her experiences tasting throughout the wine regions of France, especially Bordeaux, and I replied: "You will learn the same thing by going on the street and selling wine. Dijon will not teach you anything you do not already know. Yes, it will be fun to spend a year in school in Burgundy, but you can learn as much selling wine for me." Truth be told I was being a bit hyperbolic, but never having been one to let the truth get in the way of a good story, I did not think she would take my suggestion seriously. I continued: "I learn better doing than I do learning theory and at the end of the day, did you make the sale or not? No amount of training will make you a good salesperson. Total knowledge of the product and the industry is primordial, but the act of selling has to be in your bones."

Our discussion ended around noon and a friend had invited me for lunch. But as our conversation continued, I thought, "Why not get Diana invited?" It was a guy's lunch and I knew Diana would liven up things a bit. I called my friend Philippe Franchois and said "J'ai une fille ici au cuverie, est-ce que tu derange si je l'invite a dejeuner?" (I have girl here at the winery, would it bother you if she came to lunch?) He mumbled a not very pleased "pourquoi pas" (why not) and said "à toute de suite" (see you soon).

Philippe, as he and all our friends will admit, is a bon vivant. He led the life of the quintessential French bachelor in his Provençal-style house on the Montagne de Beaune, a hillside with magnificent views of the vines, town, and the Alps in the distance. Phillipe is also a dragueur (a real flirt) and loves the ladies. When we arrived and before he saw Diana, I could sense he was pas en forme (not in form) because his lunch plans that were always meticulously planned were now amiss with another person to serve. As much as I adore and admire virtually everything about French entertaining, spontaneity is not part of it, it is simply not in the French's DNA. As soon as he saw Diana, his eyes lit up, he rushed to greet her, kissed her hand and then both cheeks and whispered to me, "Tu ne m'as pas dit qu'elle était une jolie blonde." (You did not tell me she is a pretty blonde.) Having sensed in a very short time that Diana could more than hold her own with men, French or otherwise, she charmed him with her impeccable French, knowledge of food and tasting blind—she correctly identified the vintage of a red Burgundy we were drinking. She made quite an impression.

After a very leisurely lunch and several matches of pétanques, I excused myself at 4:30 pm because I was on duty to preform pump overs and punch downs on the fermenting wines (pumping the fermenting pinot noir juice over the solids or punching down the solids and putting them in suspension in the liquid). Diana asked if she could come with me and observe. As we descended to the winery, my interest in her went beyond career counseling. When the phone rang in the middle of my tasks, I asked Diana to take it and she said that it was Philippe and that we had all been invited by a winemaker friend for a drink at their Paulée (harvest party).

"That would be fun, they are a great family, his wife is a great cook." I paused, then quipped, "ask Phillipe about the wife's singing." Diana looked puzzled, and as Phillipe spoke, she gave me mixed messages. First, it was a "you kidding me?" look. Then an enthusiastic "Let's go." I shrugged my shoulders, said, "we warned you," and off we went.

By the time we arrived, the harvest team was well lubed. But before we stepped on their path, Philippe and I reminded Diana that when the hostess begins to sing and dance to Edith Piaf's "Je ne Regrette Rien" or "La Vie en Rose" over and over again, it is time to go. Phillipe said "Peut-être que tu peux simuler un mal de tête ou que vous avez vraiment besoin de rentrer chez vous, parce ce que vous commencez une tournée demain, sinon cela ne fera que devenir laid." (Perhaps you can fain a headache or that you really need to get home because you start a tour tomorrow, otherwise it is only going to get ugly.) She looked at us as though we were mad. We smiled, gave each other a knowing look, and settled in for the show. It took us a few glasses of wine to approach the level of the congregation, but a jolly time it was until Edith's trilling voice came over the speaker. I looked at Phillipe knowingly and he at me as our hostess was sashaying around the room. Diana watched the dancing queen with wine in her right hand and her left ready to hit the replay button on the CD player. She gave me a "What have I gotten myself into?" look and I felt whole again.

Welcome to my world, chérie. I hope you like it and I hope you will stay around for more.

Alex and Fabrice (on the forklift) moving a wine barrel at the first winery in 2000.
Courtesy of Carol Alpers

Alex and Michael Lafarge during the 2000 Hospices de Beaune weekend.
Courtesy of Carol Alpers

Alex, harvest 2000. His mother commented "You look happy."
Courtesy of Carol Alpers

Alex in the early days at his first winery.
Courtesy of the author

Alex at his newly purchased Bâtard-Montrachet vineyard in 2011.
Courtesy of Jon Wyand

Cara Schwindt during the 2009 harvest in her polka dot boots.
Courtesy of the author

Alex overseeing the harvest on a rainy day.
Courtesy of the author

Alex passing a bucket of ripened grapes in 2011.
Courtesy of the author

Alex checking on the quality of the grapes.
Courtesy of the author

Alex and Diana relax at their home in Orches, 2013.
Courtesy of Jon Wyland

La cave du Paradis (Heaven's Cellar) in the new winery.
Courtesy of the author

Barrel aging was easier in this 18th-century cellar.
Courtesy of the author

Preparing a wine tasting at Domain Alex Gambal.
Courtesy of the author

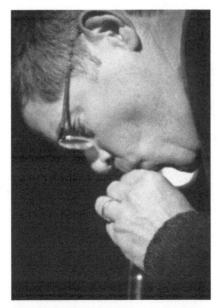

Sucking the wine out of a barrel with a pipette.
Courtesy of the author

Alex and Diana on vacation in the Sudan, 2017.
Courtesy of the author

PART 5

La Saison Verte

The Growing Season

May to July is the busiest period of the year for a vigneron. The sun is high in the northern sky and the vines can grow as much as a foot in one day. It is difficult to imagine the speed at which vines can grow. The sun can be scorching and the grapes are at risk of sunburn. Then there are periods of drought. Irrigation is forbidden for appellation wines in France. When the rain falls, the vines soak it up. But sometimes the rain is so continuous it drenches the vines and they require countless treatments against mildew. The threat of hail is always on our minds.

Chapter 18

Thrilla à Manille
(A Thriller in Manilla)

It happened fast. In the winter of 2003, Nancy and I separated; in the spring, Diana and I moved in together to an apartment in Boston's North End off Hanover Street. The historic neighborhood was known as the site of the "One if by land, Two if by sea" maxim and Paul Revere's Ride. It was a continuation of the period of financial instability of the company, and de facto personal unrest, but I seemed to be getting used to it and became a pro at rolling over my lines of credit. Companies such as AOL were issuing credit cards with a $5,000 to $15,000 line at a zero-interest rate for ninety days. After changing my address several times and applying for as many as five cards at a time, I knew I had reached a tipping point. When I started to get rejected for new cards, I felt rather pathetic. On a parallel track, I was still trying to unload the 1998 bottles for cash, which I used to pay off the cards: not an HBS or Wharton finance-approved method, but effective. I actually invoiced and declared my direct sales on my taxes, which I had no license to do, but I was aided by The Big Brown Truck acting as our company mule (in case you didn't know, it is illegal to ship alcohol, even through private shipping like UPS and FedEx). This was about as close as I got to a modern version of rum running (Joe Kennedy would have been proud). But the adage "You do what you got to do when you need cash" was my mode of operation. Based on the history of the North End, Boston's Italian neighborhood, Diana fit right in as she was often seen flagging down UPS trucks on Hanover Street to ship wine.

In the evening over a glass of wine, she would shake her head and say, "This is sooo not like me. I always color between the lines. I never do anything illegal. I am going to get busted!"

I would reply: "It will be fine, if you get busted our buddy Bill Brown will work it out with the licensing board: no problem. What I worry about are my bankers in the US and France who always want their money back." Call me crazy but for some reason my rationalizations did not quell Diana's anxiety.

Our apartment, affectingly called the Boston Bled, not to be confused with the Beaune Bled, which I still kept, coincided with and backed up Boston's twenty-year Big Dig, the most expensive and controversial highway projects in the US at the time. (Described by a local Bostonian as "A great idea if you consider $22 billion to bury a couple of miles of road one hundred and twenty feet in the ground so you do not see the traffic a good deal.") Our ground-floor hole in the wall was not without its benefits as the apartment window had a lovely view of the ventilation and exhaust tower of the Callahan Tunnel connecting Boston with Logan International Airport as well as our neighbor's ornamental cabbages. The radioactive cabbages somehow grew in the tower's shadow and we swore they glowed at night.

From a practical standpoint, our apartment was conveniently located near Boston's T (train service), the airport for flights back to France and cheap pizza, allowing me to manage my separation without too much drama. It also allowed us to easily sell wine up and down the East Coast, which Diana began to do more and more as I continued to work on permanent sources of financing both in the US and in France. In retrospect, it was clear that the economy was getting stronger and through a series of confluences, which I will discuss in detail later, sales and margins were improving though it was not evident at the time.

After a torridly hot summer of canicule (a heatwave that claimed thousands of lives all over France and Europe), the August harvest (only completed with dry ice) was one of the earliest on record. We returned to Burgundy and a series of sales events in Paris, Holland, and Belgium, terminating with the Hospices de Beaune auction the third weekend in November. Our event in Belgium was scheduled for November 11, Armistice Day (the end of World War I, which is especially important to France as they lost nearly two million citizens). With this in mind, remembering that day holidays are stretched into four-day weekends in Europe, we met my friends Hans and Marion, and learned more about the Dutch.

The Dutch are a great people. Besides being some of the most attractive in Europe and in the world (I still cannot get over how tall all the men and women are), they are bright, speak English better than most Americans, and are simply great fun to be around. They are also, as far as I am concerned, the champions of Europe's le camping car (trailer or RV) and can be seen migrating south each May to France, Italy, or Spain to escape the damp gray clutch of the Dutch delta, absorb the sun, and maintain their perpetual tans.

Having said that, there is a reason New Yorkers are New Yorkers and the Dutch are the Dutch. "Going Dutch" originated in Holland. It is not that the Dutch are cheap, it is that they question the cost of everything and everything is open to negotiation. It is totally logical that this thriftiness in their DNA permeates their worldview when you consider they have no natural resources, grow their flowers on reclaimed ocean beds and their cows occupy small squares of arable land above the various dikes (I have never figured out where they keep all of their RVs when they go home).

Besides having created a country Le Pays-Bas (Low Country) by literally pumping water 24/7 and holding back the sea (for God's sake will New Orleans please hire the Dutch), they literally invented trade. No one better represents the Dutch than my late and dear friend Hans Brockstart. Hans was six-foot-five, blond, and bronze. His wife, Marion, was nearly six-feet tall. I met Hans and Marion in the spring of 1996 at an Easter lunch that continued for many hours due to the late arrival of the caterer. As Hans would later say "We had many beautiful bottles, and we were boozing it up."

Hans, was a natural salesman as are many of the Dutch. He spoke four languages, understood a couple of others, was an incessant worker, and was crazy about Burgundy. He worked for Columbia Records for thirty years as a special collection salesman and did a bit of wine sales and brokering in Holland on the side, so he took an early interest in helping me.

I knew I could count on Hans and Marion when, in May of 1998, just after bottling my fifteen hundred cases of 1996s, I was desperate to air ship fifty cases to Boston for my first Wine Festival on Nantucket. I had to sell something.

I had flown back to France from Boston with six hundred labels that I had quickly printed up on peel and paste paper to stick on the bottles. My label with its unique font (Volgare, a 15th-century Florentine script) was developed by a focus group of five boozy Boston lawyer friends. At our local hangout, a bar in the financial district, I bought everyone drinks, put four different labels on the counter and said pick the one you like best. Everyone pointed to my favorite, the Volgare, which endured.

Labels in hand, the wine was being picked up the next day. I, of course, was hosting a last-minute wine tasting with a potential California importer. With my stress level rising, the dynamic Dutch duo stepped in. When I came up from the cellar an hour later, the floor was covered with the back peels of six hundred labels and the bottles were labeled, boxed, sealed, and on a pallet ready to go. It was then I knew these two were real friends.

Hans began to represent my wines in Holland. To illustrate Dutch negotiating skills, Hans arranged in late 1999 a meeting with an old-line importer in Rotterdam known as Hosman Vins. I flew in on an early flight to Amsterdam and we drove down directly to the importer. He was located along one of the old quays that had been gentrified but still smelled of trade. I carried in a case of 1997s to a room that appeared as though nothing had changed for four hundred years: large dark timbers held up the ancient wood floors stacked with cases of wine from all over the world. The proprietor, Ruben Hosman, was about my age, 42, courteous, showed an interest and a real enthusiasm for my wines. Mixing English, Dutch, and French, he said, "Tres bien, le Meursault et très typique et le Volnay est trés soyez" (The Meursault is typical Meursault and the Volnay is very silky)—a true compliment.

As we tasted through the wines finishing with a Grand Cru, my Clos Vougeot, I turned to Hans and said, "Wow, Hans, the wines are singing today, they are delicious." I was worried because often the travel makes wine taste at best average.

Hans, in his baritone voice said "Beautiful, they are beautiful. I want to booze it up all afternoon." The wines had rallied, as had I, from our overnight flight and all was right with the world. As we ended the tasting, discussed price, terms, and shipping, I asked Ruben what he thought of my wines (I did not ask if he liked them).

"The wines are not very good, they are too thin, not expressive, and are too expensive." Crestfallen, actually sick to our stomachs and not understanding because the wines had tasted great, we mournfully drove back to Hans's home in The Hague with our tails between our legs. As we sat in his garden and started to drink the rest of the wines in the bottles we questioned our very wine souls.

"Hans, were my wines that bad? Are we terrible tasters? Are my wines not what the market wants?"

Hans mumbled, "I do not understand, I do not know, I am getting depressed."

After several glasses of wine and good nap in the sunshine, Hans's son returned from his job as a car salesman. He asked us how our tasting and sales call went. We dejectedly related our day. Hans's son gave us a look.

"You dummies. Dad, of course Ruben said the wines were bad, not what he wants to buy, and cost too much. We (the Dutch) are trained to say from the time we can talk. Everything is always too expensive or not good enough. Have you forgotten, everything in Holland is a negotiation!"

"Jesus Christ, of course it was part of his negotiation." Even the master salesman had succumbed to the romance of Burgundy.

Over subsequent years, I sold wine to several importers all over the world who would play the same game. I would immediately counter the degradation of my product. "The price is the price. No discounts." Then I'd say seriously: "I do not negotiate with terrorists." This would always get a good chuckle, especially from our Dutch importer, perpetually red faced and smiling as he ran in all directions doing the wine pick-ups himself in Beaune to save on shipping.

A unique aspect of the wine business in Europe, specifically France and its Belgian cousin, are its wine fairs. These seasonal shows are ways for small producers to sell direct to the consumer in an intimate and non-internet way because direct sales from the winery, de rigueur in America, are still a small percentage of fine wine sales in France. Many small family domaines often block out six to eight weekends a year for these sometimes very modest fairs that dot France and Belgium. My Burgundy neighbors, for example, load up their truck or pull a trailer with fifty-to-one-hundred cases to sell direct over the course of weekends throughout the fall. Being on your feet all day is a tough way to sell, but it has it rewards in that there are no receivables and often you get to stay and dine with customers who have become like family. This said, driving from Beaune to Honfleur in Normandy to Tours in the Loire the next weekend, Lille in the north, and a final fair in Alsace can test any winemaker's energy and love of the profession.

Direct sales are no truer than in Belgium because of Burgundy's historical ties to the "low countries." It is important to remember that the Dukes of Burgundy were the eight-hundred-pound gorilla of the Middle Ages and controlled the eastern half of France from Orange in the south to Flanders in the north with parts as far east as Switzerland. (The King of France, a weak figurehead, was king in name only and only controlled Paris, aka the "Ile de France," which survived because of its walls. It was the Burgundians who captured Joan of Arc and turned her over to the English with a wink from the king when he could not meet the Burgundian's ransom.)

The Belgians love Burgundy, drink a great deal of our wine but paradoxically do not spend a great deal of money on it. There is the Ardennes and Flanders, traditionally more farming and then north sections near Brussels, Antwerp, and especially Bruges, where our mercantile system arguably was born and developed. As crass as the following observation seems if you drive through the two regions, you will easily recognize the sense of wealth as seen in the Mercedes, BMW, and Audi dealerships in the north. Annual southern pilgrimages are made to Burgundy to fill their trunks with the year's allocation. But each year, the buying spree diminishes as the older buyers' cellars are full or they die off with the younger generation no longer seeing Burgundy the bargain it was (remember, direct pricing is often the export price plus just the VAT).

With this as context, Diana and I were invited to a wine fair in Belgium, not far from Namur, on the way to Brussels and in the Ardennes in the World War II Battle of the Bulge region. The fair had a good number of participants already signed up. It was in an area with a decent population of historical importance, not that far from Burgundy. I thought, heck, they have invited us, the entry fee is small, this is their first year, let's get in on the ground floor and see if the fair has good long-term prospects. Diana and I loaded up the truck, now a larger, used Renault with a green stripe down the side that was lucky to hit fifty-five miles an hour with a pallet of wine, and schlepped up to Belgium for a Thursday-to-Monday afternoon fair.

One of the most extraordinary Christmas experiences I have ever had was the time I visited Hans in early December. If you have never read or heard the writer and comedian David Sedaris read his "Six to Eight Black Men," you must. (It is better described as six to eight elf helpers in black face accompanying Santa on his Dutch house visits.) The Dutch, along with areas of northern France and Belgium, traditionally celebrate St. Nicholas's birthday December 5 with this strange ritual. I witnessed and nearly wrecked my car as I rubbernecked on my first trip to Holland to watch Santa live on a barge, moving briskly down the canals with his six to eight helpers. On arriving at Hans's house, still astonished, I asked, "Hans, am I crazy or did I just see Santa on a barge with six to eight helpers in blackface?" He became 6 years old again and replied excitedly "of course, fantastic, you saw St. Nicholas. When I was a boy, my friends and I were scared to death of St. Nicholas, were afraid that if we had not been good, Santa would throw us in his sack and take us back to Spain."

I am still not certain if it was my imagination or if I really saw this big Dutchman's eyes dart a bit and his lip quiver as he remembered his childhood. I wanted to take all of Hans's 6-foot-5 frame, 250 pounds, and 70 years in my arms and with a hug say, "Santa is not going to take you to Spain, you have been a good boy."

He began to recover, thought a minute, and then said with his deep laugh, "On second thought, I wish St. Nichols would now put me in his bag and take me away from here in the winter to Spain or even better, Majorca."

With comedy, geography, history, and, not to mention, wine sales now in perspective, we arrived at the fair literally in the middle of a beet field that had small mountains of just-harvested sugar beets ready to be processed. The sugar beets oozed a certain irony that was not lost on me because it was the beets that were refined into inexpensive sugar that allowed chaptalization to occur and to a certain extent made Burgundy what it is today.

Chaptalization is the process of adding sugar to unfermented grape must in order to increase the alcohol content during fermentation. The technique is named after its developer, the French chemist Jean-Antoine-Claude Chaptal, and despite the polemics over whether it is a positive or negative process, chaptalization has helped wines to be made more consistently from year to year. (With climate change and much riper grapes, chaptalization is now a rare occurrence. Our current worry is over ripeness not under ripeness.)

Namur, as I discovered retrospectively, is a strategic crossroad that the Celts, Romans, and, yes, Germans, coveted because of its two rivers and its heights. The German invasion in World War I made quick work of the region despite Belgium's heroic defense and was held by the Germans throughout the war. In World War II, it was still equally strategic but its history was a bit more complicated by the initial German invasion, the subsequent 1944 Allied advance, the German return volley in December 1944 (Battle of the Bulge), and the final Allied advance.

We pulled up to an ancient, charming farm building with typically Belgian whitewashed stonewalls and think "so far, so good" as other suppliers are unloading. We enter to register and meet the organizer who is quite young, maybe 26 or 27 years old. With a beer in his hand that he clumsily spills, he says, "Hello, let me see—oh, your stand is over there." We look over at what will become our five-day visit to Purgatory. Belgium, after Germany in quantity and actually in quality, is the beer capital of the world. But when you arrive at a wine festival and the organizer has a frosty cold one in his hand that still has its foam from a nearby tap, it is 10:30 am, and it is clear this is not his first brew, it does create pause and wonder. After touring the room to get its temperature and speaking with other winemakers, we two innocents and perpetually optimistic Americans said to each other, this could be a pretty good gig. Thus was our first step into a wine fair that was its own version of the Black Hole of Calcutta.

Namur is an attractive town with an old quarter where we found a cozy hotel built into the medieval city walls. Within walking distance was Brasserie Francois, a large brasserie with high ceilings from the Belle Epoch period. The food was typical fare with, of course, the omnipresent Moules Frites. Moules (mussels) are plentiful and cheap in Belgium and the Belgians first paired them with frites (fries) because they are often credited with originating the French fry. The lore goes that they were so fond of fried fish that when the River Meuse froze up in the late 18th century, they started frying potatoes. The wine list at the Brasserie was solid with a large selection of Burgundy, but I soon realized that most of the wines on the list were special bottlings for the restaurant from Belgium négociants. This tradition of literally buying the barrels, now in bulk in cisterns, and shipping the wine for bottling to Belgium is now a miniscule part of the trade. But it was a large hub of sales for Burgundy until the late 1970s.

Hans and Marion drove down from Holland to give us a hand, really a mental health break with our sales. We thought their willingness to travel and that the fair's proximity to Brussels and even the Dutch border might entice day trippers to fill up their trunk with Burgundy. Wrong. There was a reason the Celts, Romans, and Germans made a path for Namur and it was not for their gold. Day after day, bottle after bottle was opened, poured, our story told, the customers loving Burgundy and saying: "Oh you have the best wines here" but nary buying a bottle. I do not think we sold three cases the whole weekend. As I mentally tallied the costs of meals, hotels (we paid for Hans and Marion's costs), fuel (back and forth in the truck), general boredom, and opening and burning case after case of wine, I was getting despondent.

I will not make you relive our pain, but the fair was a bust. It had, at best, a trickle of visitors the whole weekend. At any of our large tastings, we found that the timing of the crowd is based on l'heure d'apéro, or happy hour. Hans had a more succinct way of describing it: "They come to booze it up before boozing it up some more at lunch or dinner on your nickel."

We had been lured into a continuous four-day undercard event with no hope of escape. Most of our fellow winemakers made sound wines and several were a very good value for money, but the operative translation for value here was cheap. We were trying to sell Audis to the wrong crowd. But we learned slowly that our geography was off.

As Sunday rolled into Monday, we thought of bailing but felt we had made a commitment and kept hoping for one big sale or to find a serious importer in Belgium. Monday afternoon we began to repack the truck with the fifty cases we had schlepped up minus the few bottles we sold and the multiple cases our fans drank. Taking inventory was horribly painful before we limped south, 325 miles in our 55-miles-an-hour-max truck.

We arrived late at the office in Beaune and Nathalie, my office manager, all 5 feet and 100 pounds of her, charming when she wanted to be but not to be messed with, was waiting to record our sales with pencil in hand. No matter what is sold in or from France paper is involved because it is about the money, in this case, taxes. Not to bore you too much with the technical aspects, but we were allowed to transport the wine to Belgium under a special permit that let us take the wine out of France as an export (not paying tax, in essence still in bond in my cellar) yet once sold in Belgium, it would be sold as though the sale had been in France so France could collect the 20 percent tax (VAT). You have got to love the French. They get their taxes coming and going. This is controlled by a simple thing, the bottle's capsule. A bottle sold in France has a green rendering of Marianne (a female allegory of liberty), France's national symbol along with my winery's license number that proves the tax has been paid. We are required to buy the capsules and must account for them because if audited, we have to show a coherent inventory of bottles sold and tax collected.

What about tasting bottles, samples? There is a general accepted number in Burgundy of two percent for tasting samples and personal consumption but what about at tastings where sales are few and consumption is high? What are acceptable levels and what to do about all the green capsules that were used when we opened the bottles for tasting? With this as context, my office manager begins to pepper me with questions about how many bottles we burned through in a tone of "what the hell were you all doing all weekend, just boozing it up?"

At this point, as all good cowardly males do, especially entrepreneurs, I shrugged my shoulders and said "Je ne sais pas, c'était comme ça." (I do not know, it just happened.) I looked to Diana for help. "Tu parles mieux que moi et tu peux mieux expliquer les choses que se sont passées." (Diana, you speak better French than me, you explain what happened.) Nathalie was not amused, in fact, she was scary when she was on a tear. I could see she was getting wound up, so, yes, I wimped out.

Diana (my sweet dear Dee) began to calmly explain in her perfect French what I just tried to explain in my bad French. Nathalie then starts throwing out numbers, percentages, even calls the special custom's officer in Belgium who is handling the fair and begins to drill Diana like Johnny Cochrane did defending OJ and the glove. Dee, no wallflower herself, leaned in and began to go over the numbers, adopted the posture, and gave it right back. The volume began to rise on both sides of the desk and I, being the smart male that I am, retreated to my desk in the corner of the office, making myself scarce and hoping that one of the two will back down. It is interesting to note that in my experience with males, short of coming to blows, one will back down or later file a lawsuit, not females. I kept thinking this is going to end soon, but my lesson in female psychology continued for way too long. I looked over trying to get their attention, perhaps call a time out, but thought better of getting too close.

At some point, whether it was from sheer exhaustion or a pause to begin another round of the same argument, I saw my chance to get between the two of them and say "let's look at this tomorrow." I hustled Diana out the door thinking Nathalie was busting my chops over some stupid bad paperwork and perhaps 200 to 300 euros of phantom taxes. Later, I realized we blew close to 2,000 euros on expenses while the banks were still breathing down my neck and we wasted a long weekend. Nathalie, of course, was focused on the bloody paperwork not being perfect. I thought my head was going to explode.

To this day, I am reminded of those icy stares from Nathalie and admonitions from Diana about my retreat to my inner male. We named our Belgium weekend in code: Namur (pronounced by Hans as "Naaammmuuurrr"). Our response to the code was wine fairs—don't; depend on great friends—do; selling in Belgium—hard; organizers with beer in hand—flee.

Upon our return to Beaune from the land of Moules Frites, Diana debriefed my American friend and our neighbor Denis Toner. Denis smiled, chuckled, and shrieked as if he was announcing at the Muhammad Ali vs. Joe Frazier III fight of the same name: "It was a Thrilla in Manilla!"

Chapter 19

Un Âge d'Or (A Golden Age)

My editors insisted on two things: do not make this a wine textbook and do not skip around too much with the timeline. At the risk of another admonishment, I think this is a great place to step back and look at the qualitive leaps Burgundy has made in a generation.

I arrived in Burgundy in May 1993 and business was tough. Humans have an unbelievable ability to forget pain and the past (women: think childbirth) but Burgundy was in a difficult position. It was not teetering, but compared to today's quality that has led to unprecedented demand, you cannot even begin to compare the two. Winemakers such as Mugnier, Lafon, Roulot, Mortet, the Dugats, de Vogue, Cathiard, Engel, Bize, Mugneret-Gibourg, and the Gros cousins among others were known but followed only by a small universe of groupies. Even the iconic domaines of the time, Domaine de la Romanée Conti, Leroy, and Leflaive would admit today that even though their wines sold well, the less celebrated years sales could be slow.

If you think I exaggerate, I can tell you that in 1993 and 1994 while working with Becky Wasserman, we were finding and selling multiple cases of wine from the celebrated 1990 vintage as well as older vintages from many of the above domaines while receiving new allocations. Burgundy, as any other product, was not immune to economic fluctuations, dock strikes, and general market weakness. Many of the names noted above were in the early years of their careers. Fred Mugnier took back most of the family vines in 1984, Dominique Lafon was in the process of doing the same over a period of years and Jean-Marc Roulot had recently returned to the family domaine. As a point in fact, in 1994 we put together a special promotion to help sell Frederic Mugnier's 1989 Musigny old vines because he still had one hundred cases in stock that he was happy to list at 250 francs (38.00 euros) and even give a ten-percent discount for ten or more cases. In 1995, we needed to do a similar volume discount for Philippe Engel's 1992 Clos Vougeot at 150 francs (22.90 euros). To borrow a word from former chair of the Federal Reserve

Alan Greenspan, I do not believe you can describe the above as a case of "exuberance" as today these wines sell for eight to ten times more.

Twenty-five years ago, Burgundy was a hit or miss proposition. Now there are burgundophiles who say I exaggerate, but if I get some of the same vignerons mentioned above in a quiet moment after several glasses or bottles, the truth comes out. Things were tough, the wines were at times sketchy and wines made by winemakers and viticulturists from throughout the world were good and only getting better. To be fair, this was recognized and to the Burgundian's credit, change was happening.

As you might have guessed, I love history and that's why I add this bit of history—to understand the rapid but recent improvement of wine throughout the world and its effect on quality and consumption.

For almost 100,000 years, the only tool used by humans was a stone hand axe. Sometime after 100,000 years, plus or minus, man figured out if you attached it to a piece of wood you created a fulcrum. It could transfer energy, making progress and life easier. Winemakers are a bit like the maker of a stone axe—slow learners. The earliest remnants of wine were thought to be discovered in the site of Hajji Firuz Tepe, in the northern Zagros Mountains of Iran. The wine dated back to the Neolithic period (8500–4000 B.C.) and carbon dating confirmed the wine was from sometime between 5400–5000 B.C. Most likely stone age men by chance left some fruit in a bowl or a jar far too long. It softened, the juice ran, and it started to ferment, someone then took a sip and got happy. What we call wine was invented, but it took a long time to get from there to here. In all seriousness the first wines, which is a very loose term, were fermented juice—mostly grapes that had very little color but were sweet because they contained unfermented sugar, spoiled quickly and were mixed with water. The idea that wine was a lifestyle choice was a bit in the distant future. But by the time of the Egyptians, who have wonderful tomb paintings of grape growing, harvest, and winemaking, to the Greek and Roman bacchanalian festivals that at times even I would consider a bit excessive, wine became part of many people's daily diet.

Diet is an important word because it was an integral part of the ancient world's life, not because it was a fad, like eating raw food, keto, or wanting to simply make everyone else around you miserable, but because it was good for your health. The water was filthy, hygiene was poor except for the rich (who could take baths), and food spoiled and could only be preserved with salt, which was expensive and rare. Fresh fruit and vegetables were only available in season, if at all. Peasants, part of the mob, were dependent on the bread doled out to keep them from rioting between the circuses (my ancient history professor friends will find this a fair summary).

Wine was a multi-purpose product. It was used as a way to dilute and clean up the water, provide calories and vitamins, used as a religious prop, and finally, as a bit of mother's little helper to keep the mob at bay. It was espe-cially useful as a reward to soldiers for a well-fought battle and was always part of the soldiers' daily ration. In Greek mythology as well as in the real-life excess, it was the maker of legends. Alexander the Great in a drunken stupor once ran a pike through one of his generals who was his best friend. Suffice it to say what the ancients were drinking was a bit dodgy, but considering the fact that Legionnaires often signed up for twenty years, it did make the time go by more quickly.

At some point, it is believed that the Romans started cleaning out their empty wine amphora with sulfur because even they thought it might be wise to put new wine in clean containers. They discovered that the wine tasted better and lasted longer. Thus, the dreaded sulfite was born. Evidently this is nothing new. In small doses, few people have a direct allergic reaction to it. Sulfur, sulfur dioxide (SO_2), to be exact, is nothing more than an acid that acts as a disinfectant, anti-oxidant, and preservative. This is why we put lemon juice on freshly cut apples for our Thanksgiving pie. It keeps them from turn-ing brown. Lemon juice is an acid and an anti-oxidant, like sulfur. Today, there are often more sulfites in your average fruit juice and dried fruits than in most fine wines. My nose is sensitive enough to sulfur that I can smell it in Fig Newtons.

As with most things, too little or too much sulfur is bad. I firmly believe that most medical/allergic reactions to wine are not related to sulfur, but rather the ingredients in wines. I can assure you that the big guys in the indus-try are scared that one day they will be required to name their ingredients and they will fight revealing the chemicals, additives, coloring, and flavorings to the death. Did you ever wonder why some of the big scoring wines, or even everyday inexpensive wines made en mass taste virtually the same each and every year? The Gambal label would list: grapes, sulfur, sugar in some years (chapatalization), and bentonite, a natural earth product for fining: less chemicals than in your typical fruit juice.

With sulfur, chemistry was born. But not much changed until our friend Louis Pasteur discovered bacteria and how to control it by killing it. Because of Dr. Pasteur, the rest is history: medicine leapt forward, pharmaceuticals were developed to extend our lives, the food and beverage industries devel-oped cleaner products (thank you Sinclair Lewis for revealing the appalling conditions of the meat industry in *The Jungle*), all the while wine's improve-ments were slow to change due to its intrinsic small scale and regional roots.

The wine that we know today is a new product, not to be confused with one hundred and fifty years ago or even fifty years ago. Wine in the mid-1800s was a hit-or-miss proposition. The aforementioned cleanliness or bacterial issue was just beginning to be quantified though chemical analysis to improve fermentation and hygiene, and those who even thought of themselves as winemakers had little knowledge and little control of the process through temperature control and hard science. If harvest was early and hot, fermentations could begin in a matter of hours and the grapes could ferment out within days, producing a light wine, almost rosé in color. They were called L'œil-de-perdrix (a partridge's eye). If harvest was in October or even November, the cold grapes would macerate, soaking on their skins for as much as several weeks before starting to ferment. Here, the color and tannins in the grape's skins would leach into the juice, producing a dark, tannic, and age-worthy wine. The former wine needs to be drunk within twelve months and the latter might not be ready to drink for thirty or forty years because of its massive and often rustic tannins. The art and science of vinification was in its infancy.

Henri Boillot of Volnay and his brother, Jean-Marc, are the heirs to a great winemaking family who related a great example of this to me. Once Henri poured me a taste of his father's 1959 Volnay 1er Cru Les Caillerets. Now this is a great vineyard, a great vintage, and a wine from a legendary vigneron. Henri recounted how in 1959, as in most years, his father would bring in the grapes, throw them in the wooden tanks (cuves), go out and bring in more grapes, and let the fermentations just happen. It was hot in 1959 and the grapes picked after mid-morning arrived very warm, causing rapid fermentations (the yeasts love heat). In the 1950s, there were virtually no cooling systems and not until the late '60s and early '70s did some control of a wine's fermentation become common. While technology has created new efficacies, there are still a handful of true minimalists I consider magicians. Often they make what nature gives them. Henri's father did just that, brewing his juice by throwing in some ice and covering it with straw to cool it down. His fermentation was done in seven days and his wine was sublime: sweet and rich with smooth, soft tannins. Today, many winemakers who are described as interventionists, do not see their fermentations start until after seven days. The wine was sublime, sweet, rich, with smooth, soft tannins, and virtually no color.

Wine was a part of European's diets as seen in the stable consumption levels from 1850 until about 1900, even with replanting the vineyards after phyloxera that killed the vines in France. Phylloxera, a vine disease, killed the "Old World" vines from about 1860 to 1890. It was transported by an aphid-like insect to Europe on wild American vine rootstocks. (Think of the long vines that as a kid you would find dangling in the forest and try to act like Tarzan. These are wild North American vines.)

French consumption began to grow around 1901 and spiked in 1915 to 190 bottles for every man, woman, and child per year. It is not too hard to believe in the context of World War I and soldier's rations but in 1926, at the height of the '20s, it was still 181 bottles for every man woman and child per year. Please note that this was not the 15 percent alcohol fruit bombs produced in the US today, but low alcohol wine, often watered down, that nonetheless suggests a culture as distant from us today as the first North American explorers. There was even talk of building a pipeline, a "wineline" from the Languedoc (south of France "The Midi") to transport wine directly to Paris for its needs. Every aspect of society from farm living to city dwelling was different, although this was also the beginnings of what we today call supermarkets as transportation access to fresh farm products improved.

Concurrent with the growth of microbiology, improvements in hygiene and winemaking as a scientific profession, the green revolution of the mid- to late 20th century arrived. I am not an ideologue on this point, but science has allowed for humans to pretty much eliminate famine in the 20th and into the 21st century. There is plenty of food. It is getting it to where it needs to go and making it affordable that creates food insecurity. The plethora of food can be credited to three principal things: fertilizers, plant treatments, and plant husbandry (selection). After the two world wars, there was a need to do something with all the nitrogen and potassium that had been used to make munitions. The problem became a solution through the development of and use of chemical fertilizers. Rather than having to ship phosphates from the Caribbean or South America, or spread massive amounts of animal or poultry manure, we could now apply what we needed though less expensive and easier-to-apply treatments (spray applications). I have a wonderful poster from the 1930s in my home office showing two healthy, happy farmers harvesting excessively large bunches of grapes and extolling the virtues of engrais, or chemical fertilizers.

Along with the use of fertilizers, the development of spray treatments combated mildew. Now Burgundy does not seem humid when you compare it to Washington, DC, or St. Louis in the summer, but it is a humid environment that is perfect for growing all kinds of nasty molds, spores, and fungi (mushrooms) on the vines. I once was explaining Burgundy's climate to a dermatologist and he said, "Oh, you have athlete's foot on the vines." What drives rose lovers crazy is a fungus that in minor cases causes spots on the leaves and, in serious cases, can cause browning and wilting, thus greatly reducing photosynthesis. Oidium (powdery mildew) is even nastier. It is a fungus that grows on the grape bunches and chardonnay is especially susceptible to it. Oidium appears as a silvery/gray powder on the vines and, if untreated, reduces ripening and can leave a metallic taste in the wine. In the worst-case scenario, it will eat up the grapes. Worse than athletes' foot, I compare it to trench foot, a horrible affliction World War I soldiers developed on their feet because of constant water, humidity, and muck in the trenches. Although red grapes are less prone to oidium because their skins are thicker, in 2012, a year of horrific mildew and oidium, entire vineyards dissolved.

Finally, through plant husbandry—not GMO (Genetic Modified Organisms)—but cross breeding of grapes and rootstocks you can literally pick out of a catalogue the type of fruit you want and what you wish to grow it on. If this sounds somehow insidious, it is not and has been going on since prehistoric man first adapted a wild pea or a wild grain into a plant that produced enough fruit or grains to eat, to store, and to replant. Today, whether it is grapes for the table, grapes for wine, or other fruits on an artisanal or industrial scale, plants are chosen from a nursery or from existing plants by mass selection (selection massale) to meet the grower's specific needs: taste, yields, resistance to disease, etc. They are then grafted onto a rootstock that has been developed to thrive in different soils and climates.

To understand the fruit/rootstock relationship, imagine a NASCAR race car. The chassis and engine are the rootstock and can be a Chevy, Dodge, or Ford with different motors and suspensions that are adapted to the driver, the team, the track, and financial objectives. The body, the grafted fruit, can be a Camaro, Charger, or Mustang. It is attached to the chassis like a grafted fruit to the rootstock. Both are needed to run the race.

After fifty plus years of these standard practices, the baby boomers of Burgundy, who began to work with their families in the late 1970s and 1980s, began to observe their vines and wines. Today, the questions they pose are evident as everyone now has catchphrases on their lips: organic, biodynamic, lower yields, sustainable agriculture, natural food, natural wines, the list goes on. In 1990, these issues were just beginning to bubble up because a cadre of boomers and a few older winemakers who had never succumbed to new methods began to question what they had learned from their fathers and school and reexamined what their grandfathers and great grandfathers had practiced.

The term that took hold in the early 1990s, and one that I was taught at wine school (1996 to 1997) was called Lutte Raisonnée (reasoned management). Lutter, literally meaning combat or to wrestle. The concept, so simple, was that we will observe and make judgments based on what we see rather than a systematic, or an inflexible systems approach. We can only imagine the hand ringing, consternation from the industry and the older generations. From this came steps into organic viticulture and finally biodynamics (the use of a lunar calendar for planting, picking, and homeopathic treatments) based on the logic that if we have living soils, we will have healthier plants, better tasting grapes, and more interesting tasting wines.

As I acquired more and more vines from 2005 through 2015, I practiced organic and biodynamic farming. To be honest with you, one of my pet peeves are the terms natural, organic, bio and biodynamic, etc., with few if any universal standards that have meaning. Too often, these words are simply marketing gimmicks to make us feel better about what we are buying and consuming. There are at least three groups in Europe that certify growers as biodynamic, none with the same standards. This is why I was certified bio (organic), even though I practiced biodynamic (wholistic) farming. It was simply not worth the trouble or expense to obtain the extra label. Organic standards in the EU are significantly more stringent than those of the US. Finally, when we speak of biodynamic farming, we are principally following the *Farmer's Almanac*. This is based on a lunar calendar: there are days for planting, harvesting, plowing, treating and pruning, etc. Said treatments are homeopathic in nature rather than systemic. In other words, not spraying chemicals that literally go into the plant and treat from the inside out as antibiotics do, but on the surface of the vines as a preventative treatment.

This decision was not difficult to make. I believed that I would have a much better product because the approach placed the soil and plant's health in direct relation with the wine and the manifestation of the site's character (terroir). It was not a political statement to save the earth, worship the moon, or that organic wines are somehow morally superior. I simply wanted to make the greatest wine possible.

These changes have ushered in what many wine commentators called the Golden Age of Burgundy. The number of domaines that continue to reach sublime heights with their wines increases each year as another generation renews the gene pool and raises the standards of domaines that, to be kind, were either too lazy or did not care. Coupled with even better wines from lower appellations (more difficult to farm sites), there is more and better Burgundy than ever before. Some vintages in my generation that could have been a washout because of the weather have produced delicious age-worthy wines that surprise the critics, consumers, and even us, the winemakers.

With all the good changes, you would think costs would be lower, we could make wine more naturally, with less chemicals, fewer treatments, and there would be as much wine as before. But as I have gotten older, I believe in a few things and one of them is the law of unintended consequences. The vineyards are healthier, the plants more alive, the grapes more flavorful, and the wines, even in a difficult growing year delicious, but farming costs are higher and yields are much lower; therefore, we need to sell the wines for more. Some can, most cannot. This is what we call a problem.

Chapter 20

Les Affaires (Deals)

Confession: I am a deal junkie. I love real estate and real estate guys love doing deals. They always have and they always will. There is an inherent rush, adrenaline and fear to buying and selling property. I do not mention winning because in truth, most real estate guys are more fearful of losing than they are happy winning. The hunt is the thrill. I am sure it is in the human prehistoric hunter/gatherer DNA, or perhaps it can be linked to some childhood trauma that we can now blame on our parents or an incident at school for which, in this day and age, we could get compensated. After a deal is done, there is always a letdown that is anticlimactic.

God knows I grew up around the dinner table where all we talked about was business. My father's deals and his Eastern European roots made us all half expect the Cossacks or some other marauding eastern horde to descend at any moment not so much killing us, because then we would have no worries, but burning our garages, buildings, and houses and taking everything we own. It is no wonder I still toss and turn at night. Come to think of it, the banks in the early '90s were the horde's equivalent because we all signed joint and severally.

When my grandmother Babee, aka Babushka, my father, and his two sisters got together "Honest to God"—the Russian Orthodox version of the Irish's "Jesus, Joseph, Mary Mother of God," and the Jewish "oy vey"—was repeated about every three minutes whether it had to do with the kids, politics, church, or especially what was in the oven. Once in the spring when my children were young, Babee visited our home in Virginia as part of her annual family pass around. Our neighbors' cultivated azaleas were eight feet high at the peak of the azalea season. They were one of a kind and magnificent. I said "Look Babee, aren't they beautiful?"

She responded in an elongated Russian funeral dirge, "They'll be dead soon."

Thank Jesus and Mary that my mother kept us rooted in a Scottish/German/ North Carolinian Calvinist optimism (trust me, Calvinists are optimists compared to the Russians). This balanced out the self-destructive tendencies of the Eastern Front at my house. Overall, it made a pretty good mix and I must say it was always interesting when the two grandmothers got together. I saw them operate in the kitchen together as two ethnic bastions—Babee with her halupki (stuffed cabbages), pierogi with sour cream, and boiled kielbasa with horseradish mayonnaise, and Grammy Ruth with the best fried chicken, mashed potatoes, gravy from real drippings, and strawberry shortcake. All this deliciousness made sense to me as an adult when I learned that far from being full Russian, we are at least half Polish with roots in North Carolina. Grammy, who loved to shop at Lord & Taylor, always had a new dress on, never once splattering grease on herself. Babee wore an apron.

These ethnic origins, sweet-smelling flavors, and my "Episcopal light" boys school St. Stephen's, in large measure, were my foundation. There were about fifty students in each class from 3rd grade to high school with a daily rigor of academics, sports for three seasons, the occasional theology class that really was more of a skull session for the teachers to find out what was going on, and, finally, a weekly twenty-minute chapel service. The greatest threat of my childhood was our school disciplinarian James D. Osuna (J.D.O.), a formidable teacher of ancient and modern European history whose 8th-grade notebook and study methods caused fear, loathing, and tears. He instilled in us a disciplined system of thought and analysis that my friends and I still use today. We boys would have often preferred Catholic school damnation and a nun's scorn than being on the wrong side of J.D.O.

Despite my best efforts, I stumbled through 6th-grade Latin, nearly failed 7th-grade math, and managed to get through chemistry, thanks the kindness of Dr. Nugent. I will never forget the humiliation delivered by one coach who told me I was never going to make it. These tough lessons, balanced by my wrestling coach's encouragement, helped develop my inner drive, determination, and a fear of failure. It was not just a fear of failure, but really a fear of letting down my parents, my peers, and my own expectations that motivated me.

When my start up in 2005 was sufficiently solvent, at least to suppliers, banks, and customers, I began to seriously look at acquiring land in order to have a guaranteed supply chain. I did not buy an existing estate or a winery, rather started by buying small bits of vines. To understand how I came to own vines, it is important to understand why I needed to. For the first five years of the business, my primary focus was buying the best grapes that I could afford, which were in good supply at prices that were relatively stable. Beginning in 2003 and 2004, I saw the need to become a domaine. Although my supply chain was solid, I began to see changes in Burgundy's landscape as suppliers were more and more reluctant to part with their grapes and were putting more wine in bottles themselves. There were a growing number of small boutique wineries such as mine and many of my suppliers were also beginning to buy in wine from their friends and neighbors. Unless I created my own supply chain, small operations like mine and others like mine would simply become a bank or factoring service for small growers. We were feeding the beast that would devour us.

Ecosystem is an overused phrase, but I think it aptly describes Burgundy's economic system—one that is based on different actors that are interdependent: independent but dependent on one another. We begin with the vigneron or farmer who grows grapes, makes wine, and might sell grapes or wine in barrel. The négociants are wineries who supplement their vineyards by buying grapes or wine in barrels from the vignerons. They sign a standard industry contract with the growers promising payment over nine months from harvest. If the grower needs cashflow, he can go to his bank and borrow against the contract/future payments. These are the basic financial building blocks of agriculture in Burgundy and are still the common blocks of agriculture in US farming, where local lending institutions, originally mutual, state, or local cooperative banks, provide credit to the farmers.

Domaine bottling is historically a new practice and even today there are a plethora of small growers who sell their grapes or wine under this standard industry contract that in general specifies that the buyer pays in three installments over nine months from harvest. The price is fixed the following spring, normally at the end of March or April, and is based on an average of all the contract prices that are registered with the local grower and négociant association.

You make the contract before harvest, assume some of harvest's costs, and pay two-thirds of the balance before you even know the final price of the grapes not to even mention the quality of the wine? Yes, indeed, welcome to Burgundy. Add in the fact you often you have to pay taxes on the grapes, as much as seven percent, and you can understand why I describe the business as being a banker.

How are prices set? You could say we read tea leaves but more accurately it could be described as a far-too-cozy system between the large wine houses. I describe it as a papal bull: orders come down from above, you follow or you are excommunicated. Until about 2009 or 2010, the large wine houses and several of the principal brokers would meet in the early spring and set the average prices for all the appellations. Not to make this sound too nefarious or like a cartel, but there was a real push/pull between the large wine companies and the growers. Both ultimately knew, however reluctantly, that their respective financial welfare was tied together and prices that reflected the market were set. In the spring around Easter, the courtiers (brokers) would visit me and they would begin to tell me le tendance (the tendencies of the prices). I would have lunch with one of the major actors and get told what the prices were going to be. Invariably, even in the age of the internet, a typed sheet with all the prices per appellation would be handed or faxed to me.

Until recently, the system generally worked, but it often resulted in large swings in the bulk prices of grapes, leaving the bottle price stable, one of Burgundy's great paradoxes. Wholesale or bulk prices reflected the economy of the moment, not eighteen to twenty-four months later when the vintage was in bottle and was being sold. A great-quality vintage could tumble in price simply because the small winemakers had no money to buy, like what happened in 2005 and 2006, two terrific red vintages, or spike in price like what happened in 1998 and 2008, more difficult red vintages. (Starting in the 2020s, a system based on a three-year average price became more common. This has been the case for generations in Champagne and eliminates the highs and the lows and provides for a more stable cash flow for both buyer and seller.)

These changes only accelerated from 2010 to 2019 because of small harvests, a strong economy, a popular product, more vignerons bottling their own wine and many of the same vignerons now buying in grapes. The market became a mix of the traditional big houses with their large contracts, small specialty winemakers, such as myself and the aforementioned new growers. From 2005 to 2015, there were at least three to four hundred new winemakers, most of whom were existing vignerons who formed a parallel company that allowed them to buy grapes. Many of these negoç, as we like to call them, were modest in scope, purchasing on average of thirty barrels worth of grapes, representing 300,000 cases of wine, a conservative estimate but enormous for Burgundy's size. With expansion of existing vineyards not possible, more demand and small harvests, Burgundy's demand/supply balance changed, forcing grape prices into the stratosphere. One might forgive me for using a baseball or *Moneyball* analogy and say we are now in the age of free agency.

This was the transformative landscape I was operating in when I made my first attempt at buying vines and becoming a vigneron. In 2004, I had the opportunity to take over and lease a small domaine in Chassagne-Montrachet with some decent but not great vineyards. In retrospect, it would have been a good first toehold but I was beginning renovations on our new winery and in the middle of my divorce. When my team performed a mutiny by saying, "This is too much," I passed.

The next summer, safely moved into my new winery, I purchased from a retiring vigneron whose daughter was a friend at the wine school, four acres of Bourgogne Pinot Noir and Bourgogne Chardonnay in Volnay. This is the lowest classification of our vineyards. What I affectionately called the "cheap bleacher seats," cost around $100,000. By setting up a vineyard infrastructure or laboratory, I could learn about operations and costs with limited downside.

I knew that if I could break even on the farming and make a bit from the wine, then it would be an inexpensive business school. More importantly, I set up a corporate vineyard infrastructure that would allow me to add more prestigious vineyards in the future.

The most important feature about this small transaction was its personal relationship nature. It was the first of many opportunities I took advantage of by being sur place (a local). Fundamentally, real estate is a local business and the locals will always have an advantage. This is especially true with agriculture and in Burgundy, where there is a small universe of players and no for sale signs. These players or gatekeepers can be accountants, an occasional consultant that specializes in estate planning sales but most importantly, notaires. A notaire is a French lawyer for whom we have no real counterpart in the States. They can be more aptly described as a family priest, a rabbi or a family's consigliere. They literally know where everything is buried and often who brought the shovel. They are primarily real estate attorneys who do much more than real estate sales and closings: transmissions, or structuring inheritances, selling real estate as a broker and even divorces. They are not litigators but managers of the various Code Civils, France's code that lays out in excruciating detail what you are allowed to do. Thus, notaires who specialize in vineyards are experts in the agricultural code that spells out what someone can do vis-à-vis leases, sales, droits (a person's rights)—you cannot imagine how profound this concept is—all the while trying to keep the transaction's taxes as low as possible. Throw in usually one or two recalcitrant third cousins that have an insignificant stake in the deal but feel cheated because of how the will was managed three generations before and you can see why transactions are few and far between. Finally, notaires tend to have their sons or daughters follow them into the firm and they, in turn, follow the sons and daughters of the same clients. It is best to visualize this as a multi-generational local business that is centered on managing a family's business affairs and keeping family peace. Perhaps in another life I would become a notaire because this is something I already was familiar with given my dad's business.

My friend and principal notaire, Martine Thomas Crolet, is chatty, funny, knows everyone, but, most importantly, knows how the game is played. She will take you out to the vines to show you a parcel, put on her boots and slog through the muck while perfectly dressed, kind of like how my grammy cooked chicken. Martine is one of my heroes because she started as a secretary in a notaire's office with one of Beaune's legendary deal makers who spotted her talent, went to law school at night, and eventually took over and built up her own practice. Simply put, pulling yourself up by your bootstraps is rarely done in France. Martine lives in Meursault and is married to Meursault's mayor, so she sees deals flow. Early on, I decided to stay close to her.

One of the most iconic events of the summer is the 14th of July in France, Bastille Day. Each town and village throughout France has its own fête. Meursault's is especially splendid, held in the town square with simple picnic tables, a bandstand with many of the Meursault winemakers playing traditional French dance songs. The barbecue is prepared and served alternating years by the tennis and the rugby clubs. Old-fashioned multi-colored light bulbs are strung randomly between the trees under which couples dance.

In 2007, my old friend Russell from high school visited with his family and joined us for the party. We arrived a bit late so I said to Russell to go grab a table while I got some wine and ordered the food. Bottles of Bourgogne Chardonnay, Meursault, and Bourgogne Rouge were plopped on the tables along with plastic cups, perhaps a plate of smoked salmon, real fries, merguez sausages, hamburgers, and hot dogs. The little kids got in line for the non-stop waffles, the older boys threw firecrackers, the girls flirted with the boys who were oblivious while parents drank, talked, and laughed.

By chance, the table Russell selected was next to the politician's table, where Beaune's mayor, some of the city council, Martine and her husband, Denis, sat. At any public gathering in France, it is common for the two principle political parties to make a showing with their entourages each moving en masse like an amoeba, through the crowds, shaking hands, making small talk, and trying to get seen by the cameras. I knew everyone at the table and immediately shook hands all around and did the bises (kisses on the cheeks) with the ladies. I chatted a moment with Martine and decided this was the perfect time to give her the business.

In a light-hearted and good-natured way, I asked: "Martine, comment se fait-il que la seule chose que vous me proposiez à chaque lancer est un groupe de chiens, allez, envoyez-moi quelque chose de bien!" (Martine, how come the only deals you ever throw my way are a bunch of dogs, come on, send me something good!)

The next morning, I received an email from Martine that said, "Appelez Maître Goujon, il a de Puligny-Montrachet disponible." (Call Master Goujon—her old boss—he has some Puligny-Montrachet vines available.)

I met Maître Goujon (master, as in professional) the next afternoon. Wiry in build, slightly hunched, he looked me in the eye and told me clearly what was going on. "J'ai des vignes disponibles. Mon client M. Andre Morey, un vigneron en retraité Il a donner ses vignes a ses filles qui veulent les vendre." (I have some vines available, am representing Mr. Andre Morey, a retired vigneron who gave his land to his two daughters and now they want to sell.) They have 0.68 of a hectare (1.7 acres) of Puligny-Montrachet Les Grands

Champs and Les Petits Grands Champs (just below the 1er Cru Clavaillon), a treasure of chardonnay vines. "Voici est le prix, mais il y a des confiances qu'il faut je partager avec vous. Est-ce ça que vous intéresse?" (This is the price, but there are some confidences I must share with you. Are you interested?)

Knowing a bit how deals do and do not work in Burgundy, there is one word you say: yes. This is not the US, where you beat each other up, get everything decided, and then sign. Here, the règles (rules) are to say yes first and then work out the details (I learned this lesson the hard way in a later opportunity). What were the details? Was the site a toxic dump or Jimmy Hoffa's final resting place? It was worse: under a lease to their cousins.

Lease is a four-letter word in France if you are an owner of a vineyard and want to sell. Cousins in business, another major hurdle. A standard vineyard lease is for eighteen years with an automatic right of two, nine-year renewals for the tenant. By law, an owner cannot kick out a tenant at the end of the lease and replace them with a different tenant for any reason unless the owner plans to take the land back and farm it. In order to exploit, aka be a vigneron, you have to have at least 1.5 hectares (3.7 acres) under control and need to have been trained at wine school or you need to present a plausible case you are capable of running a domaine. If the owner wants to sell his property, the tenant has a right of first refusal to purchase the property and in any case the lease transfers to the new owner under the same conditions mentioned. It is thus logical that absentee owners, many of whom never set foot in their vineyards, have leases that renew automatically for generations.

When I give tastings, I am often asked "Why don't you just buy up a bunch of these small parcels and consolidate them? I answer the above question with a hypothetical: suppose that your great aunt emigrated from France to the US generations ago and left you 0.20 of a hectare (about 0.5 acre) of a vineyard in Beaune. You get a check for 2,300 euros a year in rent from a distant cousin you have never met, the land is worth 250,000 to 300,000 euros, and you want to sell. You call me and ask if I want to buy it? If for some reason I was already making wine or had a parcel nearby, big ifs, I would say yes but then I would ask:

1. Who is the tenant?
2. How many years are left on the lease?
3. Does the tenant want to buy the vineyard (knowing that he has a right of first refusal to purchase it)?

In most cases, the tenant will want to buy the land or if the lease was just resigned, he will say no, I am happy to keep leasing the land for say sixteen more years and you, the owner, are stuck with a less valuable and totally encumbered vineyard.

There is however one exception to death by leasing: a 25-year lease, which in this case was ending. The land at the end of a 25-year lease is libre (free) and the tenant does not have a right of renewal or of first refusal to buy. Because so many vineyards are encumbered, when an unencumbered parcel does become available, it is gold, worth a premium and all the hoops it takes to jump through to purchase. In the case of the Puligny, the daughters were benefiting from the foresight of their mother who, when their father retired, insisted that they enter into the longer-lease term, knowing that both parents could be deceased at its termination and did not want to leave their children with an encumbered and therefore highly discounted asset.

The late Monsieur Morey, diminutive in size but with a deep voice, retired when he was 60 and was still active in his 80s carving wooden St. Vincent statues (the patron saint and protector of winemakers). Carved grape bunches from pulled-up vines adorned his atelier. The image of a Burgundian vigneron, he was quick to laugh and charming with the twinkle of a natural born salesman in his eye. Morey told me that his wife, who had passed some years before the transaction, "a gérér le quotedien du domaine et bien sûr c'etait elle gardait une trace de l'argent, et généralement en liquide" (ran the day-to-day affairs of the domaine and of course kept track of the money, generally in cash). He chuckled, "Ouais (pronounced WAY), c'est moi qui ai faisais le vin et planté les vignes mais elle s'occupait de l'argent." (Yes, it was me who made the wine and built the vineyards, but it was she that watched the cash.) "C'etait son idee de faire un bail de pendant 25 ans. Elle a aucune confiance dans nos cousins à lacher le bail si nos filles voudraient vendre." (It was her idea to make a 25-year lease. She had no confidence in our cousins releasing the lease if our daughters wanted to sell.)

It bears repeating that Morey and other vignerons who began with virtually nothing but a horse-drawn wagon in the late 1940s, built their domaines with pure labor when domaine-bottled Burgundy was the exception. He told me how he and another vigneron cleared the Merusault Narvaux vineyard of its stones with a wagon full in the morning and another in the afternoon drawn by horses in order to plant vines. He also explained to me that after the Second World War, the French government put in place a tax system called a "forfait," a set tax paid based on production volume, not the value of the product. If you produced 5,000 liters of wine, you were taxed at a very low rate on the bulk amount and not the value of the wine. This system assumed that you were selling your wine or grapes directly to a négociant at a very low value. Enterprising growers who made good wine and told a good story (wizards in the wineries and charming salesmen), would be taxed at, say, five cents per liter if sold in bulk to a négociant at seventy-five cents per liter and the same five cents per liter if he bottled the wine and sold it in bottle (750 ml) at three euros. This allowed enterprising growers to have the ability to literally and legally store money in their mattresses and in turn buy and develop more vineyards. Monsieur Morey and others acquired fabulous vineyards when the value of their bulk product and land were low to negligible. I have a good friend, a vigneron in Meursault whose father operated for years under this system, and at his death, the family had to sell off in bulk literally hundreds of thousands of bottles that his father had squirreled away for years for his retirement. In the mid-1990s, I saw the stocks on several occasions and it was astounding to see full pallets of wine from vintages as far back as the early 1960s with some 1975 Meursault wine still in tank. The late English wine writer Clive Coates tells a story about Monsieur Andre Ramonet buying his parcel of Montrachet, the most coveted white wine vineyard in the world, in 1978 with cash he was literally pulling from his overalls. This is the ultimate example of this system.

The lease ended November 15, 2007, and the Morey's were concerned that their cousins who were leasing the land would try to sabotage the sale even though they had no holdover rights. Within four days of meeting Goujon and Morey, I drew up a one-page letter agreement that promised I would keep my mouth shut until November and they in turn promised to sell the property to me. The letter was nothing more than a promise to write a contract for sale five months later. I gave them a check for ten percent of the purchase price, which the attorney promised to hold and only put in escrow until a contract was signed. This is a true story and is one of the major reasons I love this place, because so much of Burgundy is based on trust and the reputation of one another that is sealed with a handshake. I knew of the reputation of Maître Goujon whose son was my banker, and he in turn checked me out via his son to make sure I could be discrete and reliable. If I had not made good on my word, I would have been persona non grata, shunned for all future purchases. Was it risky to sign a non-binding letter to promise to do something? In the States, it would be not worth the paper it was written on and the parcel would have been marketed to higher bidders, but in Burgundy it was written in blood.

When we went to Morey's home in Meursault for closing, we sat around his small table, where he ate and watched his beloved soccer matches. His daughters, Martine, Goujon, and Diana were there as we passed around the papers for each to sign. It was just before noon, and before we all went to lunch, it was time for an apéro. Mr. Morey went down to his cellar and brought up a bottle of 1976 Puligny-Montrachet. Chuckling as he came into his kitchen nook he asked me "Ca va aller?" (Is this going to work?)

"Bien ŝur!" I responded. Mr. Morey took out a kitchen knife and with the skill of a neurosurgeon cut off the capsule, looked at the top of the cork

approvingly, and then gently twisted his corkscrew that had seen better days
into the cork.

He handed me the bottle and said "Tu le fais, tu as le propriétaire maintenant." (You do it, you are the owner now.)

I did not miss a beat when I took the bottle, but thought "for God's sake do
not screw up now." Gently but firmly, I pulled out the cork. It was as fresh and
pristine as it was at bottling. "Monsieur Morey, c'est vous qui avez fait le vin,
vous devez le servir." (Mr. Morey, you made the wine. It is you who should
serve it.) Mr. Morey smiled approvingly at me, poured each of us each a glass
of Burgundian gold, and offered us his favorite elephant ear sugar cookies.
We toasted to each other's health and success.

Chapter 21

Hiroshi rêve de la Bourgogne (Hiroshi Dreams of Burgundy)

The question "Why did you choose Burgundy?" is best answered by putting the question in the context of the love affair the Japanese have with Burgundy.

It was in Japan that the answer to the question became clear. I had not chosen Burgundy; Burgundy had chosen me. You cannot choose to be something you are not unless a great many forces and circumstances align. The key is to be open and curious. The process is not linear and is fraught with risk. I have become a much more patient person because Burgundy rewards patience and thoughtfulness. This is difficult for most Americans to accept because we see the possibilities, the rationalities of scale, the power of action, and the need to be in perpetual motion. Burgundy is the antithesis of this credo. If you want to succeed in this corner of pinot noir and chardonnay, you have to put action in a drawer. It is work of a different sort.

My French friends ask in Cartesian perplexity "Why would you want to leave America and come to Burgundy?" With the Japanese, I got the sense that the same question is more existential. They too would love to move to Burgundy, want to understand why and how. Actually, many have. Several Japanese women I know have married Frenchmen, it has some superb Japanese restaurants, and there's even a Japanese winemaker in the Côte d'Or.

It is a relationship that is complex, passionate, and ongoing. During my time in Burgundy, it seemed that the Japanese became the Burgundian's twin. Except for a handful of equally intense Americans, I believe that on a per capita basis, the Japanese know more about the nuances of Burgundy, its producers, and its vineyards than any other people. They breathe in Burgundy. Ironically, it is not a surprise if you look at Asian culture. It too is also steeped in tradition, ceremony, and nuance.

Our first trip to Asia to sell my wine belied these nuances. It was a whirlwind from Singapore to Tokyo in December 2005. A relocated Taiwanese American from California told me that Singapore is Asia 101. A wine's smell is part of its language and culture, and in Singapore I could smell new Asia: a clean, perfectly run metropolis that takes pride in the fact that everything is meticulous.

Vietnam and Hanoi were just opening and the old colonial city was still there but you could see and feel the beginning of demolition, construction, and modernization (aka capitalism) in the old communist capital. It was a bit surreal because thirty years earlier as a senior in high school I saw the fall of South Vietnam on television. To be present where the conflict originated and observe the wave of modernity was a cruel irony of the sadness of the conflict. While the tastings were a mix of ex-patriots and up and coming Vietnamese shoppers, my sense was there would be a good middle class for me to sell to but it was in its embryonic form.

I had dinner with the hotel sommelier, and I am sure a couple of internal agents, in the old market. We drank longneck Budweisers as vegetables and fish were cooked to order on small gas burners. After dinner, I had one of the most amazing experiences of my life as I hung on to the back of the sommelier's motor scooter. He weaved and bobbed though the unimaginably dense and joyous Hanoi streets just after Vietnam won the Southeast Asia soccer championship.

Seoul, a massive city that seems to have been copied from a set of plans from a US city planner from the 1970s, spans two sides of a river that is the Mississippi in dimensions. Diana and I spent days visiting restaurants and retail stores where a middle class striving to move forward came in and out of Starbucks and Korean kimchi joints. My importer's new hire, a 20-something named Mr. Lee, spoke a bit of English but he did not know one end of a bottle of wine from the other. To make matters worse, he drove with his left foot on the brake and right foot on the accelerator. It was three days of non-stop whiplash and motion sickness as we sped from one side of Seoul to the other, a Burbank-to-Orange County sized jaunt with similar traffic. Diana pleaded with me to take the wheel. I was glad to have a witness to the madness. People, mostly women, were out shopping on Saturdays because their work week had recently been trimmed from six to five days a week so they were easily lured into the wine shops where we held our tastings. Korea grew to be one of my larger Asian customers.

In Japan, my importer prepared a head-spinning five-day schedule that was incomprehensible. Appointments at one end of the island followed the next day by appointments at the other end. I will never forget landing in Sapporo and taking a two-hour bus ride to the city center with my salesman and translator, a journey that stopped at every 7-Eleven discharging and picking up passengers (total distance about thirty miles). Within ten minutes of arriving at the hotel and I was expected to begin a tasting and the bla bla. The tasting was approximately forty-five minutes, the dinner another hour, and then a return the next morning on the same bus line to the Tokyo airport. As we boarded the plane, I asked the salesman in exasperation "Do you realize we could have rented a car more cheaply, not wasted time on the bus, and we would have only had a short forty-minute car drive?" The response, "it was not planned."

We took bullet trains to all corners of the Tokyo metro area, which is in reality most of central Japan. Stops were frequent and we walked through every shopping center to get to the importer's numerous retail outlets. The owner, a dynamic and creative woman, had a concept of selling wine on the Japanese version of QVC TV. She was planning at first to sell low alcohol wine not made from grapes, akin to a wine cooler, to get people to drink wine. I have a vision of her doing a homespun Japanese version of Bartles & Jaymes (the wine cooler guys) in a kimono.

Peculiar to her tastings was a *The Price Is Right* moment after we tasted and discussed each wine. She asked her customers as Bob Barker would ask the audience, "Now how much do you think this Vosne Romanée costs? 15,000 yen? 10,000 yen, no, just 7,500 yen!" It was if she flashed the audience a "clap now" sign because her comments were always followed by "ohs, ahs" and forced applause. I thought maybe this was how they sell in Japan, but as the visit progressed it dawned on me, this ain't normal. This truly was one of the most surreal experiences of my business life and was repeated with even more vigor two years later in their new upscale store in one of Ginza's chicest retail buildings. After our second trip to Japan where Diana and I were hauled around late into the night as walking, talking props, we found ourselves another more conventional importer. Honestly, we thought, there's a lot to say for boring sales calls.

It was back in Burgundy where we met our #1 fan, Hiroshi Tsuda, a self-proclaimed "salary man." Hiroshi's annual trip to the Hospices de Beaune began Friday with a 4:00 pm sharp arrival chez Gambal and a thoughtfully and perfectly wrapped gift. Hiroshi only speaks tiny amounts of French and English but his eyes always speak for him as does his infectious enthusiasm for one of the biggest weekends in Beaune.

La Ventes des Vins, The Hospices de Beaune's annual wine auction, originated from one of the most glorious structures that exists from the late Middle Ages and the raison d'être of Burgundy. Over the centuries, gifts of money, buildings, and land, specifically vines, have been bequeathed to the Hospices, making it the largest owner of 1er Cru and Grand Cru vineyards in Burgundy. The Hospices has its own full-time winemaker who supervises the farming with contracted vignerons. At harvest, the grapes are transported to the hospital's winemaker at their modern winery literally next to the Beaune hospital. The wines are then auctioned off on the third Sunday of November to individual purchasers or winemakers such as myself, with all proceeds going to fund hospital operations. Imagine if your local hospital was founded 550 years ago, owned a 10,000-acre ranch and auctioned off 1,000 calves every year with the profits going to the hospital and you will begin to understand the Hospices.

In the Côte d'Or, thousands of Hiroshis from all over the world come for the auction weekend to taste, party, and knock on winery doors. Over the years, I got a bit smarter and limited my tastings to an open house on Friday afternoon so as to try to limit my sleeping in the winery in between constant cycles of washing wine glasses. To add to the performance pressure, Diana's tour company, The Hidden France, had their marquis' trip over Hospices' weekend. She drafted me as her lovely assistant as she squired clients to some of the best-tasting addresses and tables in Burgundy.

After almost a full week of partying with guests and clients, I am spent. I do not need another great bottle or great meal. In fact, I often ask "Can I please have some clear broth and sit in front of the television for a few days?" I do not want to sound like a spoiled brat, but there is a difference between being a host and a guest when your principal job description the entire year is hospitality. By the end of November, Burgundy and I am exhausted.

Hiroshi's permagrin revealed his passion for my wines. When a wine was not shown, he acted as the doting uncle who loves all his nieces and nephews and wonders, "Where is little Susie? Is she ill? Will she not be here for dinner?" His attachment to the Burgundians and their wines was fervent and his nose could always sense when you had something new to show him.

Hiroshi had some favorites, our Puligny-Montrachet, Batard-Montrachet, Beaune Greves but he especially loved Chambolle. For many years, we had our current selections displayed on a shelf in the office tasting room. After he faithfully gave his greeting, he would move with the grace of a dancer to the shelves to see what was new. This was his reconnaissance, and I am sure he was plotting to see what bottle(s) he wanted to bring back to Japan. Invariably at the end of the tasting if I did not bring out one of the children, he would ask "Can I have a little taste of the Batard-Montrachet?" His manner of asking with his thumb and index finger pressed furtively together increased the level of platitude without raising his voice, all the while looking left and right to make sure he was not overheard, because he knew that if he had other conspirators his chances of having a special taste were doomed.

After these moments, because it was rarely just one wine, he would ask in his sweetest voice, do you have a of bottle Chambolle Les Amoureuses or Chambolle Les Charmes that I can buy? The beauty of his maneuver was that he loved to have one or two bottles that no one else had in Japan. He seemed to know exactly what the importer brought in, its price, and if it was available. His end game amused me and I was happy to be his co-conspirator. As our tasting ended, he would borrow one of our bikes and I would send him off in the dark of November with more trepidation than I had when I tried to teach my children how to ride a bike. He piloted the Côte d'Or for the next five days in all types of weather as his inner childlike inquisitiveness guided him on all his annual two-week French tour de vin.

Once the Hospices weekend ends, Hiroshi heads to Paris on a crusade to discover the newest restaurants and wine bars highlighted by the five-day Salon Des Vignerons Independants (Independent Winemakers Salon) at the Paris Expo center: his wine heaven. At the Expo, more than a thousand French winemakers gather in one place and allow you to taste without guilt as they sell their wines direct. For five days in a row, he is like Ralphie from *A Christmas Story* on Christmas morning as he finds his Red Ryder BB gun.

After nearly thirty years in France, I dream in both French and English. Hiroshi speaks little French or English but he always dreams of Burgundy.

We kept a Hiroshi-like pace during our sprint in Japan. After returning to the hotel late, waking up at all hours from jet lag (décalage horaire), visiting the lobby, perhaps the gym (but more likely the bar), and trying to get ready for the next day, we were lost—physically, mentally, and sometimes emotionally. I remember the moment staring out over Tokyo harbor and thinking I am a stranger in this place where they call me gaijin (foreigner) like Bill Murray in the film *Lost in Translation*.

Lost in Translation is one of my favorite movies because of my international travel. It has so many layers, the traveler in a strange land, jet lag, the sweet relationship between Charlotte (Scarlett Johansson) and Bob (Bill Murray), the actor playing an actor. When Bob is in Japan shooting a commercial for Suntory whisky and there is a wonderful scene in the movie where the interpreter is a bit fast and loose with the translation.

Director [in long format formal Japanese, to Bob]: Mr. Bob. You are sitting quietly in your study. And then there is a bottle of Suntory whisky on top of the table. You understand, right? With wholehearted feeling, slowly, look at the camera, tenderly, and as if you are meeting old friends, say the words as if you are Bogie (Humphry Bogart) in *Casablanca*. Here's looking at you, kid—Suntory time!

Interpreter [In English, to Bob]: He wants you to turn, look in camera. OK?

Bob: . . . Is that all he said?

I felt the same about the translation and tried not to laugh or cry at a tasting as I looked out at the audience of about one hundred Japanese people were nodding in unison, except my friend Hiroshi who gave me a look that said I was getting short sheeted. Hiroshi, always the subtle and polite Japanese man, showed me pursed lips and nervous horizontal eye movements. Mind you, I do not speak a word of Japanese but after four to five days of repeating the same wine routine even a linguist blockhead like me can begin to get the sense of what is being said. Please bear in mind when I try to say arigato (thank you in Japanese), it usually comes out as aligoté the wine. After the tasting, Hiroshi, who knows as much about Burgundy and my wines as anyone, was as visibly upset in public as I imagine any Japanese person ever gets. He just kept shaking his head and plaintively said "no, no, not good translation."

The question of translations and translating is very interesting because when we translate we are not simply changing words, we are changing their collective meaning and essence. Wine tasting is the same exercise; we try to decipher the tastes and reassemble them into a meaning, but the pieces are not always interchangeable. I am always curious to really know what goes on between an interpreter and a head of state. Does the interpreter have to really understand the essence of the leader and his policy position? Where do the science of language and the art of communication intersect? I often wonder when Trump and Putin had their tete-à-tete behind closed doors in 2017, what really was said? What if a translator went rogue? Is there a backstop, a translator checking the translation? All I could do was plunge into the abyss. As long as they like the wine, bought it, and invited me back, let the translator have at it.

At one of the last events I attended in Tokyo in a very large auditorium with close to two hundred attendees on the top floor of one newest and swankiest addresses, we had a tasting seminar where I decided to try something different. This being the last night, I purposefully strayed from the script to see the reaction of the translator who was getting on my nerves. A pleasant, very bright man who spoke both English and French and went to the wine school in Burgundy, was always nervous as a cat to make sure I followed the script.

I left the stage and walked around the aisles for questions in a town hall-style, asking them up close "What do you think of the wine?" "You must come visit us in Burgundy!" It was quite fun playing my best Oprah getting the audience to open up and show me their inner Burgundy. The crowd was 80 percent women and as I approached them, they blushed, delighting in the way I was talking to them personally. They even gave me little gifts. I was quite flattered in turn and for about a nanosecond, I felt like Mick Jagger (OK, maybe Bill Murray).

I was on a roll pointing out the various vineyards on the photos, going backwards and forward in my remarks when the translator tried to cut me off. He literally grabbed the microphone and said: "We are out of time." Now this really agitated me. I grabbed the microphone back and said, "I did not fly halfway around the world to be cut short because the schedule says so." I practically got a standing ovation from the crowd and instantly fell in love with the Japanese. It was one of those moments of recognition when I entered a narrow hallway into a new culture and let the audience drive with me outside the lines.

Alex with Hiroshi and friends at the tiny wine shop in Ginza, Tokyo.
Courtesy of the author

When in Tokyo, Hiroshi was always a wonderful host, showing me off at his favorite hole-in-the-wall restaurants. The owners all had the same passion for Burgundy.

One Sunday night, I realized that Hirsohi had only one love. We sat on the doorstep of perhaps the smallest wine store in central Ginza (if not the world) and the owner cooked us a 1-star dinner on a Bunsen burner. Even though the place made a Manhattan apartment kitchen seem big, there was Burgundy everywhere in what I came to call the store on the stairs. He had Rousseau, DRC, Dujac, Mortet, Mugnier, Leroy, Meo, all the Gros family, Lafarge, d'Angerville, Gouges, Coche, and all the greatest Côte de Beaune whites. And to my great pleasure, many cases of Gambal. You literally had to walk down to the basement sideways (not so different than some caves in Burgundy). Claustrophobic does not begin to describe the space, which was cramped with 500 to 1,000 cases of wine stacked from floor to ceiling as its walls. It was the wildest retail operation I have ever seen.

To put the evening in an even more surreal perspective, you would think the streets would have been quiet, given that the shop was in a retail district and it was Sunday. But this was not the Hanoi market, it was Ginza: across the street were Louis Vuitton, Gucci, and the Chanel boutiques. The Japanese women love to shop, and on the weekends, they become Olympic champion shoppers. As we sat on the sidewalk on folding camping stools, drinking our wine, and eating multiple courses, gaggles of well-heeled perfectly coifed and quite attractive Japanese women passed by. With Hiroshi and his friends, it is always drinking not tasting. We drank God only knows how many bottles and were feeling quite happy.

I asked Hiroshi, "You tell me it's hard to find a good Japanese girlfriend, but look at all these attractive women shopping. What's wrong with them?"

Hiroshi replied nervously, "I don't know, I don't know, complicated."

I thought "What the heck?" and said, "Hiroshi, let me go up and talk to some of the women and I will introduce you. They will like you."

"No, no, no" he said with panic in his voice, "Japanese woman very complicated." Then, he warmly looked down at his glass of red burgundy, smiled, and said: "Burgundy is easier to understand."

Chapter 22

Un Loi au Citron (A Lemon Law)

I admit it, my first major purchase was way too easy. But once I dipped my toe in the pond, the water got deeper. Martine called me in early December 2007, just as I returned from an Asian trip and asked, "Alex, souvienes-toi le parcel de Chassagne je t'ai montré cet automne, est-ce tu es toujours interessé?" (Alex, do you remember the parcel of Chassagne I showed you this fall? Are you still interested in it?) I said, "Seulement le Chassagne, le parcel de Maltroie?" (Just the Chassagne, the parcel of Maltroie?) Oui, Vendu! (Sold!)

Not only was the Puligny-Montrachet easy to acquire but Martine had a file (dossier) of vines that she was selling that was a terrific collection of a former Chassagne domaine. It was far too large for me to purchase, even if I had not been buying the Puligny. But Martine, clever as she is, spread the wealth by offering parcels to me and various other vignerons in various villages. This not only kept the price from being negotiated and discounted in an en masse sale, it was also smart politics because it made multiple vignerons happy in multiple villages: no one was 100 percent happy in getting everything but no one was 100 percent unhappy in getting nothing. If a village syndicate, or growers' association, complained that they did not get enough, she would counter and say that no one got everything and if you keep complaining next time you will get nothing.

In an instant, I said yes to a half acre of Chassagne-Montrachet 1er Cru La Maltroie that would produce 100 to 125 cases of wine each year. This is a normal and perfectly average transaction in Burgundy, but anywhere else winemakers and grape growers think we are nuts with these tiny parcels. What made this even more extraordinary is that I had been making this exact wine since 1998 and it not only fit into my range of wines but also met my goal of expanding my land holdings while decreasing my dependence on outside suppliers. I called my banker who added this to the financing of the Puligny, called a few friends who contributed capital, and slapped my hands—done deal.

The purchase of the vineyards was completed in February 2008. Over the next two years as I converted the vineyards to organic farming, I learned a great deal. I did not have much of an appetite to buy more land because of the financial crisis and the ensuing drop in sales. Burgundy sales are a lagging financial indicator; I have found that Burgundy tends to enter and feel a recession late and likewise come out of it late. This is primarily due to the export nature of our product. Importers find that their newest shipment invariably arrives just as a recession is beginning thus the ability to turn on/off the spigot is delayed as is the final sale of inventory.

In May 2010, business was just beginning to revive, and I heard that a domaine in the Côte de Nuits was available. It was a collection of parcels of Vonse-Romanée, Chambolle 1er Cru, Gevrey-Chambertin, Morey St. Denis, and two postage-stamp-sized parcels of the Grand Crus Clos de La Roche and Clos St. Denis. It also included the vigneron's stocks of wine in bottle and perhaps his tractors and building, depending on tax issues. This was a transaction I call a Humpty Dumpty: you take it all, break it apart, keep what you want and put it back together, trying to limit the tax burdens that usually are transferred from the seller to the buyer.

The vigneron's wine in bottle was not bad if not great, but it was unclear what he would ultimately do with the buildings and tractors (I figured I could use them in the short term) but it was the two tiny pieces of Grand Cru paradoxically that worried me the most. One parcel was only partially planted and was literally part of a wall. Together, they would perhaps make a barrel of wine, twenty-five cases. You cannot legally combine two small lots of Grand Cru grapes and still call them by their respective names (although it has and continues to be done). Add in the difficulty of vinifying a single barrel of pinot noir. Possible, but it would push the limits of credulity to put it mildly.

The next day at lunch I agreed to the terms, no negotiating the price, and asked "Pouvez-vous me donner 24 heures pour penser aux deux minuscules parcelles de Grand Cru? J'ai besoin de penser comment les gérer?" (Can you give me 24 hours to think about the two tiny parcels of Grand Cru. I need to reflect on how I am going to manage them?) We shook hands, broke bread but I walked away forgetting the first and most important rule in Burgundy real estate transactions: he who hesitates, loses. Early the next morning the broker called me upset because the seller decided to take another deal. I was crushed, mystified as to what the other deal was but mostly angry at myself for sweating the small stuff. I made a basic tactical error and vowed never again to make the same mistake. I was correct in my analysis that the two tiny parcels were of no practical use to me, but my hesitation was misplaced. I was mistaken in not realizing how valuable those two tiny parcels would be as vineyard capital. It would have been easy to lay off or trade those two parcels to a neighbor for something I could use because the neighbor would be desirous of my asset. In retrospect I came to see how important it is to judge parcels not simply as vineyard trophies or even as a production vehicle but as players on one's team or as the deeds on the monopoly board that can be traded and used as strategic assets maximizing their utility to you while minimizing their cost. My lack of dispassion clouded my judgment; it was a hard mistake to admit. I am not talking just about the monetary value of a parcel, but the cost benefit of the trade to the coherency of my wine offerings. It is a great deal easier to trade or sell 1er Cru and Grand Cru parcels than a parcel of Bourgogne or average Village wine. The total calculus must be carefully assessed. But in this case, the kid who loved to play Monopoly blew it.

In early July, I heard both from Martine and the broker on my busted deal that they were working on a new dossier. Within two weeks, I was given a rough outline of a transaction that was rapidement (rapidly) coming together with tantalizing hints that it included some Grand Cru Bâtard-Montrachet. It would be a post-harvest transaction when vines are usually offered, with the goal that the new owner takes possession of the vines in the winter to begin pruning and other vineyard work. But I was stumped; whose Bâtard was it? I went through my records, researched online but could not figure out who's domaine this was.

In October, the sale's package was ready. It included a large parcel of Bâtard-Montrchet, with a total sales price in the 8 million euro range. Could the broker come, have me sign a non-disclosure agreement (NDA), and explain the project to me? Bien sûr. We met the next day, October 12, I signed the NDA and he returned on the 14th to hand me the sales package. It included three of the best parcels of Bâtard-Montrachet, some fantastic Puligny-Montrachet, and a bit of Chassagne-Montrachet that was being sold by one of my former wine school teachers, Philippe Brenot.

For someone who prides himself on knowing who owns what, I was dumbfounded. I remembered that Philippe made wine; many of the teachers at the wine school taught for the pleasure and made wine as their profession. But I never realized the depth of his holdings. I only remembered him as a non-stop talker and me and my classmates desperately trying to keep up with his rapid-fire lectures in between puffs on an always-lit Benson & Hedges cigarette. His enthusiasm was as infectious as his hearty smoker's laugh, and as I would come to find out, his love for a "ca-ca" or espresso coffee (I hope you are getting the picture of his energy).

Because of the Vosne-Romanée fiasco, the broker promised me that I would get the first look at his new project and he was true to his word. The next week, he took me on a tour of the vineyards, buildings, and we met Philippe. The crown jewels of the domaine were his parcels of Bâtard-Montrachet and Puligny-Montrachet Les Enseignères (which was the key to the transaction) along with some parcels of Santenay, Santenay 1er Cru, Bourgogne Pinot

Noir, and Bourgogne Chardonnay (imagine an acquisition of several super-stars and a bunch of utilitarian players, of which you have plenty). From the late 1980s until the early 2000s, Phillipe had strong US sales but he was at the point in his life, even though only five years older than me, that he wanted to downsize and make a small amount of wine for fun. He also had a brother from whom he rented half of the vineyards and with whom he did not speak. Perhaps he was simply tired of the family drama.

On the surface, the transaction seemed straightforward enough, but there were hurdles that made this deal extremely complicated.

- The vines were free (libre) of encumbering leases, but the contract required a significant lease termination fee that was paid to Philippe, not to his brother. The termination was to remain vague and was built into the price of the vines.
- I had to purchase all of his vines, the majority of which I did not want. I could lay off some of the vines before closing, but because of the short timeline, the bulk would be sold after closing, thus financing was needed.
- I had to purchase his winemaking company and its stock of 12,000 bottles, which would have serious tax consequences for me. This process of buying wine in bottle in bulk (en vrac), and not at "Two Buck Chuck" prices (a nickname for a wine in California famous for selling for less than water), is a standard négociant procedure. A new commercial label is created and the bottles are resold at a tiny profit to the grocery stores or as we say "GD" (Grand Distribution). This had never been my business.
- Approximately twenty barrels of the 2009 and the 2008 vintage that Philippe had not yet sold were to be purchased by me and resold (or bottled). The 2010s had already been sold in bulk but I was purchasing his stocks of wine that always cause tax and goodwill issues.
- A future tax bill from the Chassagne-Montrachet vines that were owned and trapped in his operating company, with no tax step-up in France.
- Two houses in Santenay: a falling-down farmhouse and a run-down duplex on the square.
- Two third-party leases that did not fit my wine profile and needed to find homes.
- A leased parcel of the very rare Chassagne-Montrachet 1er Cru Remilly that had great value but was leased directly in Philippe's name and thus not easily transferred.
- All the unknown fun and games you find when you buy a company and come to a final sale in four months for 8.65 million euros, plus closing costs of 6.2 percent and we were talking almost 13 million dollars.

Perhaps a better way to sum up this kind of transaction, in fact most transactions in Burgundy, is that the seller is trying to sell everything, including Granny's house and sometimes even with Granny in it.

Acquiring these vineyards would be a major step in the growth of my domaine and make the perception of my Domaine much more prestigious because I would be a member of the "Grand Cru Club."

Is it fair, misplaced, or outright stupid that the world of wine puts such gravitas on one vineyard and then transfers that same gravitas or halo effect on you and all of your wines? Of course, but this is no different than any other luxury product. The Bâtard-Montrachet alone would not change my life, but it would change the perception of all of my wines and create great value for me. By making Bâtard, I would be making its wines at the highest level. I would be married to it and it to me, as it is and should be in Burgundy. As we visited and I examined the assets and I tried to get a handle on the various pieces of land, their location, who owned what, third-party lease terms, bottles of wine, the two buildings and their condition, and finally the two jewels—the Bâtard and the Puligny. My mind got mushy and my heart started racing as the clock ticked. It was simply too much to process and assess in two to three hours, but I knew there was only one decision to make. "Let's make a deal" (passons un contract).

We then visited Philippe and it was as if we had never left class. He lit up cigarette after cigarette, he opened several bottles (which I would be buying), we talked about his goals, my goals, and I felt that even though the unknowns were numerous, Phillipe was excited to sell to me. Selling to an American was unheard of. Because he was my teacher and an Energizer Bunny at that, Phillipe would be an active ally in achieving closure. We shook hands and we said "let's do this." I came to find out it was huge relief for Philippe to sell to me and not to one of his local friends or colleagues. He told me after the sale that his friends were just being nice to him because they wanted his vines. I could tell this was a great disappointment and hurt him because he had spent his whole life in Burgundy and even he did not expect his friends to be so self-serving. He and his wife Christine were also glad to have the stress of selling behind them and have their life's work in the hands of someone that cared for it as much as they did.

That same afternoon, I wrote a one-page letter agreeing to the sale package that was presented to me. If the seller agreed, I would draw up a purchase and sales contract and that my offer was without any financing conditions. The promise to sell, even shorter than the one written with Mr. Morey in 2007, effectively blocked the sale to others. Blocking the sale, or more specifically entering into exclusive negotiations, is a way to tell everyone else to back off and not tamper. I am reminded of the rules of engagement or ownership in Melville's *Moby Dick*, and the chapter entitled: "Fast Fish and Loose Fish."

The rules are unwritten but clearly understood by the parties of the fishery. If they are not followed, it means chaos. It is another paradox of Burgundy that with so few sales you would expect more of a rough-and-tumble atmosphere, but because there are so few opportunities, the rules of relationship win out. Like calling fouls in a pick-up game of basketball—if you do not admit fouls or exaggerate others' fouls, you will not get an invitation to play again or, more likely, you will be leveled flat.

As bizarre as it sounds, there was no rush to sign a formal purchase and sales contract and we set the signing date for a month later during the week of the Hospices de Beaune with formal closing set for February 1, 2011. On the morning of November 18, I signed the documents for Philippe and his wife Christine's portions of the transaction. At 6:00 pm, Martine and I met Philippe's brother and wife, who had driven down from Paris, at their notaire's office in Dijon. The families communicated only through their respective notaires and as outgoing as Philippe was, his brother was the opposite: taciturn, quiet, and deferential, especially to his wife who regarded each document with the tenaciousness of a Jack Russell. What should have been a formality became a bit of an inquisition as she went over all the details and made me and Martine question if they were actually going to sign the contract.

They were recalcitrant at best. Was it just his sister-in-law's way of exerting control or was she rethinking the entire transaction? A case of seller's remorse or a venting of the family feud? As a family business survivor, I came to see it was the latter and for two hours we felt the wrath of a lifetime's perceived slights. Under this barrage of verbal and non-verbal blows, I decided to play a rope-a-dope like Mohammed Ali, absorbing the hits needed to win because I wanted to return to Beaune with a signed contract and not start over. Martine looked at me with a plaintive expression and to her credit, played along by cowering like the little sister getting reamed out by her nasty, older sibling.

Madame and Monsieur finally signed but not until the notaire read out loud the contract, each and every word, a standard procedure in France, and received verbal agreement from each participant that they understood what they were signing.

In the car, Martine looked at me and I at her and we spoke over each other in French "C'etait incroyable, c'est pas possible. Elle est une vrai salope! J'etais certain le deal éetait foutu." (It's incredible what just happened. It's not possible. She's a real bitch! I was certain the deal was dead.) Martine gasped. "Il auait un moment ou gue j'etait certain nous allions rentrer Beaune sans un deal." (There was a moment when I was certain we were going back to Beaune without a deal.) We calmed each other down on the way to see Fabrice then recounted the same cinema as he shook his head in disbelief and then they both turned to me. "Comment vont la levée de capitaux et finance-ment?" (So how is the capital and financing going?)

"Sans soucis," (No worries), I replied, "On prends un coup." (Let's get a drink.)

I called Diana, gave her the update, and drove to meet her at Denis Toner's house in Beaune. Denis, not only a great wine man but an amazing cook, was having one of his legendary pre-Hospices de Beaune dinner parties. On one such occasion, Denis got Jacques Lamelois, the well-known 3-star Burgundy chef, to give to his Ris de Veau a thumbs up and handed him a Crestor. He offered cholesterol pills regularly as after-dinner mints at his house (Jacques had recently recovered from a heart stent, Denis had multiple stents, and I entered the club in 2009).

I arrived to an assembly of friends anxious to hear if the deal was done. Diana, several winemakers, and Dorothy Hamill and her husband, who are friends from Nantucket, were thrilled to greet me and celebrated me as the first non-French owner of one the Montrachets. It was especially sweet because, as some of us of a certain age may remember, Dorothy won a gold medal at the 1976 Winter Olympics. She was as adorable in person as she was on television back then.

After an emotionally charged day, our better judgment intact, Diana and I returned home at a reasonable hour and I instantly fell asleep. When nature called at 3:30 am, I began to reflect on the day and the project with all its vari-ous permutations. The number of euros began to simmer, then boil through my cranium. The more I thought about the deal, the more agitated I became.

What the hell have I just done! I have just signed a contract for nearly 13 million dollars and I do not have the money! I do not have any of it, I have not even begun to put an investment package together, how am I going to get this done? I literally started to shake and to have a panic attack and thought if I do not complete this deal, I will be radioactive. No one will come near me. I am dead; they should just bury me now. Christ, I am having buyer's remorse before the sale is even complete. I wonder if I can get out of the contract. Is there a Lemon Law in France?

I returned to bed hesitant to wake Diana. I lay on my back, perfectly still, trying not to shake and hyperventilate and I began to talk myself off the ledge. It was clear that this was the biggest project I had ever undertaken in my life. It was also clear that my ability to raise 13 million dollars in a bit over two months was highly unlikely. However, it was possible, as I had first discussed with the broker in October that with a great deal of work I could lay off a large part of the risk and make an attainable six-million-dollar transaction. Now, where do I find six million dollars?

Chapter 23

C'est Pas Possible
(It's Not Possible)

January and February 2011 were spent raising money. In retrospect, it was a mistake to allow the months of October/November to slip by, but I reminded myself I did not have a signed deal. For Christmas, we were with Diana's family in an idyllic Vermont setting. I was under deadlines to pull various documents together with little to no FedEx service because of a three-foot blizzard. While everyone else skied, I worked the phone, calling friends, friends of friends, and clients, casting the biggest net possible for one-on-one appointments in early January.

Starting in North Carolina and finishing in Boston, with stops in Washington, New York, my frustrations with how little time I had to pull this together made my mood as nasty as the Nor'easter that stranded me.. In New York, I met up with my daughter, Alexa, who was living and working downtown across from City Hall. At the time, it was just beginning to be the livable and dynamic neighborhood it is today. We walked in the damp cold with the wind whipping up between the buildings as it only can in Manhattan and I asked her to choose a good fish restaurant. As we wandered, she said "I think this is good." It was just what I needed—a quiet, warm moment to talk to my daughter about something other than my current obsession, find out how she was, and enjoy some fresh and simply cooked fish. I asked for the wine list and saw that they had some half bottles of white Burgundy and in fact one from 2000 that as I took off my glasses to read, realized was Phillip Brenot's Puligny-Montrachet Les Enseignères!

"Oh my God, these are our vines!" Serendipity does not come close to describing wandering into a random restaurant in New York City and coming across a bottle of wine from exact vines you are about to buy in France. "Alexa, 2000 is really old for a half bottle but what the heck, let's try it." It was just what old white Burgundy can and should be. Not oxidized, still fresh but with the nuttiness and round mellow flavors great old white Chardonnay grapes give and this in a half bottle. It was so good that we ordered and drank a second half bottle. It was so good that I ordered and drank a second half bottle, seeing this serendipitous moment as a sign that the deal will close. We kept the bottles, one of which is still at my New York partner's home.

Apart from that affirming moment, I spent the trip deconstructing and reconstructing the cost benefits of this epic acquisition. In retrospect, I numbered them.

One, it was a huge positive that the vines were available and not encumbered, even if it meant being harangued by Madame. But, how was the lease termination fee to be paid? Was it an expense? Could it be capitalized? In the end, the cost went on the books as an asset and could not be expensed, which was a bummer. This was a large tax advantage to the seller and a major disadvantage to me: in other words, it was goodwill.

Two, I had to purchase all of the vines at once but was able to lay off half of the Bâtard along with the worst of the houses. I kept two of the three best parcels of Bâtard and all of the Puligny-Montrachet Les Enseignères that sold in bottle at the level of a 1er Cru, a 30 to 35 percent difference. We did not find a buyer for the Santenay 1er Cru Passetemps until after closing, so I had to personally carry the financing for two months.

Three, tax considerations greatly benefited the sellers because sellers invariably inflate or deflate values usually to the disadvantage of the buyer. For example, the Chassagne vines were owned by Phillipe's company that I was buying and in France there is no "step up" for hard assets. The seller sold the company, not the vines, and as a result the Chassagne-Montrachet vines were valued at their original cost basis, leaving me with a future capital gains bill.

And four, leaving aside the real financial loss on reselling the wine, what was even more costly were the hours necessary to dispose of it. In the spring of 2011, Burgundy was just coming out of the recession. Merchants still had plenty of wine in stock no matter what the quality of the wine for sale. In this case, even the celebrated 2009 vintage was hard to peddle. This showed how low a market can go and how profitable this trade would have been if done three years earlier or later.

I could write a good short story about the preparation and resale of the bottles because I thought that this part of the cleanup would be relatively painless. I was a salesman after all. I did not want to sell Philippe's bottles under my name, but I could use his name because I now owned his company and brand. A broker found a buyer for the first 2,500 bottles at a price that, if not exuberant, did not cause me too much anguish because we did not have to clean and label them and they were sold as is or nu (naked). When we cleaned and labeled the bottles, invariably we were short labels and that required me to get a small, expensive print run.

The day before pick up, I delivered five, 600-bottle metal cases for their transport. When I arrived at Philippe's, I saw Philippe and the broker staring at the trap door in the floor with arms folded thinking hard.

"On commence et décrivons les cas à remplir" (let's get started and descend the cases for filling), I said.

Philippe looked at me as I looked at the trap. "C'est trop petit; ca ne marche pas comme ca ici." (It is too small; it does not work like this here.) "Il faut descend le chariot par la grue et charger deux caisses de vendages par fois et pois remontrer le chariot." (You have to lower the dolly by the crane, fill two harvest boxes at a time with the bottles, and bring them up again on the dolly.) I did the math—48 bottles x 52 boxes = 26 round trips.

I let out a primordial, "Mon dieu" and Phillipe said, "Prends un de tes gars le faire." (Get one of your guys to do it.) I have no "gars, il n'y a que moi, je suis à le gar, tout la monde travaille dans les vignes" (I am the only guy available, everyone else is working in the vines). "Comme toujours je suis le petit lapin à faire le boulot" (as usual, I am the little rabbit to do the work)—we would say the shmuck.

I will spare you the gory details but we developed a system to get the bottles down from their placements that required ladders, bending, leaning, crawling, and finally lifting the filled boxes onto the dolly, centering the dolly below the hole, putting on the steel hooks as the cable and chains lifted the spinning ensemble all the while I expected the ancient wooden crane or its steel cables to snap, falling and killing me instantly. I kept a good distance back as the contraption rose but at certain times, instant death seemed preferable to moving another five thousand bottles that continued to be sold at bigger losses. I assure you I could not make this up. Finally, we then have to take the bottles out of the boxes, fill the steel cages, and then descend the dolly and boxes for another round.

Each box of 48 bottles or four cases weighed at least 75 kilos (165 pounds) so by the end of this first foray, we lifted 3,900 kilos (8,600 pounds) for each loading and unloading. My team and Diana could never understand why I was so exhausted and in a nasty mood after my bottle visits chez Philippe and why I did not hire someone to do it for me. By the time I found someone who wanted to work, made a short-term contract, and had to deal with the probability that the guy would not show up, I thought it would be faster and cheaper if I did it. I did make a note to myself: in the future double the cost and the loss on bottles bought on site because of the permanent physical and mental damage that moving 12,000 bottles inflicted.

At this point, I stopped going through the list. The house we retained in Santenay (Granny's House) turned out to be a real surprise. Early in my work career, I renovated and rented houses and I knew how to manage contractors, but more importantly, I could imagine how the finished project should look and function without an expensive designer. In other words, with some experience, you know there are only so many places walls and doors can go and never, and I mean never, try to move existing plumbing in an old house. You do not want to know what you will find and you might as well tear the house down and start from scratch. With these caveats, we combined the two small hodgepodge units into one clean guesthouse for our clients, friends, and the occasional renter.

The third-party leases found homes without any real headaches but the ability to absorb the Chassagne-Montrachet 1er Cru Rémilly proved elusive. If the lease could have been retained and assigned to me, it would have mitigated the physical and fiscal pain of the bottles. Philippe had the law on his side, but with the owner notoriously uncooperative and litigious, Philippe did not have the stomach to fight and I had no leverage.

By reducing the needed capital in half, the deal felt doable, albeit with major hurdles to achieve, there remained one small detail: how to pay for it?

My first investors gave me the confidence to pull this off. Knowing that someone actually believed in me and understood what I was trying to accomplish was a big boost. Of course, having committed capital begot other capital. And I forgave myself because panic is a natural reaction, if not at least a great motivator.

In mid-January, two weeks to closing, having told Martine and Fabrice time and again everything was on schedule, I called and said we have to put off closing until March 1. This calculation was based on a gamble that I had leverage, albeit temporarily. If they asked for money for the extension, which they did not, I could pay them and buy some time because I finally had some capital in the bank. I felt that they would say yes to my delay because they would not want to restart the process. I had one good card to play and used it, but I still had a long way to go because I had just a bit more than one-third of the investment I needed to close.

On a trip to Florida for wine sales and another round of deal selling in February, I raised another third of the capital. But trying to register the partners was proving to be a nightmare: passports, current addresses, dates of birth and where, dates and places of marriages were necessary for the partnership documents. French documentation is, if nothing else, thorough and detailed.

Back in Burgundy, papers completed, my refrain was that "the money is coming." Two weeks 'til closing, Martine had me over to her office for a come-to-Jesus conversation. "As-tu l'argent?" (Do you have the money?)

"J'ai besoin plus de temps, encore un mois." (I need more time, another month.) She replied "C'est pas possible." The French always say "pas possible" (it is not possible). This is not as in English where we think, it is not going to work, or it is proven that it will not work, or there is no way I can make that appointment, plane, etc. In French, it is different and usually means I do not want to be bothered or it is not in the regulations or I had not thought of it that way and because it is your idea, it is not possible.

Martine, however, was saying it in a way most of us can understand (aka you are out of luck and I am not hanging my neck out for you anymore). I said "I understand," but she finished by saying, "J'ai une idee, je t'appelrerai plus tard aujourd'hui." (I have an idea, I will call you later today.)

That afternoon, she called me and said that her banker, Jean-Paul, would be calling me. He called twenty minutes later and we set up a meeting (rendezvous) for the next morning, Tuesday at 11:00 am in my office. He and his colleague arrived, I explained the project, that I had raised two-thirds the equity, but I needed to fill the gap of 1,200,000 euros. "Je crois qu'on peut faire quelque chose. Je vous appelerai avant la fin de la semaine" (I think we can do something and will call you before the end of the week), he said. He called Friday morning and said "J'ai une offre pour vous que je pense est interessant." (I have an offer for you that I think is interesting.)

- 1,200,000 interest only (rates were going lower)
- Two-year term
- No penalties for paying off the loan early
- A lien, not a mortgage on the property (mortgages are expensive in France because the government charges big registration fees). This was in essence a bridge loan.
- Open business accounts with us and let us bid on your banking business
- A personal guarantee

Now, after going through the nightmare of bank guarantees in the early 1990s you would think I would say no. However, let's be honest, I had no leverage but the bank was betting on me to succeed and all I needed was time. There was zero chance of raising cheap capital in less than two weeks, but I believed that it was highly probable if not 100 percent certain I could raise the capital and pay off the loan in two years. In essence, the bank was selling me time or writing me a cheap life insurance policy while I was in the emergency room. I had a health scare, but if treated it would heal in 24 months with hardly a scar.

From the bank's perspective, they saw it as little to no risk. The transaction had a 2:1 loan ratio on some of the best vineyard real estate in the world. I had never defaulted on a loan, my business was growing, the bank needed new business, and I was guaranteeing them more business with a loan that came with the land as collateral. Some would argue that I assumed way too much risk, but I beg to differ. With the loan in place, I assumed defined risks but controlled a great deal of the equity in an asset that I was certain would

Philippe and Christine Brenot at the sale of their Batard-Montrachet vineyard.
Courtesy of the author

quickly grow in value. With a deposit check of 30,000 euros, significantly less than one percent, I had tied up 13 million dollars-worth of assets. No one asked for more money or guarantees and I for sure did not offer because I had none to give.

After any large transaction, no matter what the business, there is always talk on both extremes. Either you paid too much and got robbed or you got lucky and stole it. After the closing, I heard through a broker that one young vigneron said I was crazy to pay that price. Peeved, I replied it is always easy to say someone else pays too much when you inherit a domaine of five hectares of some of the greatest pinot noir 1er Cru and Grand Crus vineyards in the world.

Not long thereafter, I received a call from another broker saying he heard I had sold my Batard and Puligny vines to the legendary winemaker Madame Lalou Bize Leroy. Puzzled, I said I had not and what was he talking about? In fact, Bize-Leroy had purchased similarly sized parcels of Bâtard-Montrachet and Puligny-Montrachet Les Enseignères from another vigneron in Puligny-Montracht for 30 percent more than what we had paid not three months earlier.

PART 6

Véraison

Ripening

Véraison is the moment in the growing season when the grapes begin to change color. The vigneron's work in the vines is largely completed, vacations begin while we wait to prepare for harvest. A calm comes over the hills as vineyard treatments have ended and the tractors and white Sprinter vans that are ubiquitous to Burgundy and its vignerons rest idle in their garages. It is time for the leaves to do their photosynthetic work. With the cooperation of the last third of summer, we are taken to harvest.

Chapter 24

Terroir—Pouvez-vous jouer de la musique? (Terroir— Can You Play the Music?)

What is this word *terroir* (pronounced "tear-warr") that arouses contentment among the French, a knowing nod from Americans who drink its Kool-Aid but sends so many scientists, wine professors, and new world winemakers into apoplexy? Why are the French so at ease with the concept of a sense of place and its character and a few naysayers desperate to prove them wrong? Merriam-Webster gives terroir a simple explanation: "the combination of factors including soil, climate, and sunlight that gives wine grapes their distinctive character."

In 2004, I was invited to the International Pinot Noir Celebration in McMinnville, Oregon held annually at the end of July in the heart of the Willamette Valley. The celebration entails three days of eating, tasting, and drinking pinot noirs from all over the world and is about as good as it gets. Great chefs, interesting tastings, and everyone jolly about some of the freshest food in the world make it one of the greatest adult fraternity parties that still exist. Beyond the fun and games, there are a series of serious wine tastings and lectures with journalists, importers, and winemakers where we share ideas, philosophies, and answer usually very thoughtful questions while we share our wines. I was on a panel called "Le Nouveau Visage de la Bourgogne" (Burgundy's New Faces). Besides being flattered to be included with some terrific winemakers whom I call friends, it was simply great fun to have tastings on two consecutive days moderated by Allen Meadows, aka "The Burghound." Allen, a self-described former "financial type" is arguably Burgundy's most thorough wine journalist. No one else covers Burgundy with more detail and passion.

The contrast between the energy of a tasting with American winemakers, who often need to be restrained as they talk about their wines, and the French is not hard to notice. Think of a bunch of 8th graders all talking at once in the school cafeteria and you might come close to American winemaker's energy. The French, and of course there are exceptions, let their product do the talking or let others do it for them. I do not think it unfair to generalize that French winemakers, and really any French producer of a fine-quality product, are reticent to promote their distinguished products. This is especially true of the small and insular group of Burgundians who all speak a wine dialect specific to and rooted in the terroir. This is not to say they cannot and do not like to show enthusiasm and cannot explain (sell) their product but, in general, especially for my colleagues on the dais, this dialect is hard to overcome, especially while trying to speak a bit of English.

One very large exception was Jacques Lardière, a French version of Doc Brown in the film *Back to the Future,* who was Louis Jadot's winemaker for forty years. Jacques was and still is one of Burgundy's great ambassadors.

Fast forward to 2010 on Nantucket where I was the moderator, someone asked Jacques to explain the 2006 vintage. He began slowly and then stood up, hair askew: "You must understand that there were special electrical forces at work in 2006." He swirled his hands above him like electrons looking for a charge, then switched their direction. "Sometimes positive and sometimes negative. The energy of the vintage affected the fermentations, but it was all related to the grapes and the electricity at harvest." He went on for more than five minutes explaining the unseen forces that were at play at harvest. To my immediate right, the late Phillip Senard from Domaine Comte Senard in Aloxe-Corton, a dear friend who was a delight, spoke perfect English, and teased me about my wonderful French accent. "Ut oh," he chuckled as Jacques continued to rile himself up. "Il devient excité, nous serons ici un petit peu." (He is getting excited, we will be here for a bit.) He reached for a bottle of his red and poured himself a full glass to steel himself for the force of nature most akin to a tsunami: that is Jacques.

At the end of Jacques's discourse, delivered in English with a heavy French accent, there was a pregnant pause in the room. I raised my hand and said "Perhaps I can translate." In about two minutes, I explained the 2006 harvest conditions and the types of wine made and Jacques looked and me and said "That's right!"

When I do similar tastings in France, the French say my accent is charmant (charming) but I have my doubts. I am fearless and get almost as excited as Jacques showing my passion for the place and the product. Diana told me perhaps I should have a bit more fear as I get going faster and faster and my French becomes even more wacky as I pull out my visual aids of rocks, dirt, and maps. I am sure that everyone has learned a great deal, the participants perhaps not so certain, but everyone leaves with a smile on their faces.

Back at our tasting in Oregon in 2004, with a different character, I was at the far end of the dais to audience's left and next to me was Christophe Perrot Minot. In ten years, Christophe took a family domaine to the heights of quality. After the basic bla bla of each of us talking about our wines, and Allen trying to get the panel juiced up, Christophe was asked: "Mr. Minot, how do you make your wine?" Without missing a beat, he answered "C'est le terroir." (It's the terroir.) I sensed this coming and as I was not quick enough to interject a qualifying question in Christophe's ear before I put my head in my hands. I thought "Christophe, can we please forget the terroir for once and tell them how you make your freakin' wine!"

Poor Allen gazed at me with this look of, "Help me out here, buddy."

I looked at Allen and said: "Allen, perhaps I can help. Can I rephrase the question?"

Allen relaxed and gave me a look of "God, yes, please help. This tasting and discussion is going nowhere fast."

I then asked Christophe in French, "Christophe, la question est apropos des details de tes vinifications" (Christophe, the question really is can you explain to us your vinification techniques)? This is the beauty and the problem of language: context. When I explained it this way, Christophe's jumped back on board and he started giving all his winemaking techniques.

Then a follow-up question from the same person: "But isn't this the same, or about the same as all of you make your wine?" Allen looked at me and shrugged his shoulders.

I took microphone and said "let me translate—what Christophe is saying is that, yes, we all make wine in about the same way. Of course, some people pick earlier, some later, some destem their grapes 100 percent, while others use stems, some do a long pre-fermentation while others do not; the permutations are endless but the point is why are the wines different? Because of the terroir: the place, the site's exposition, and a million tiny factors that make up the micro climate. Once we deconstruct the wine, literally look how it is made, we can begin to understand that the magic is not in its making but in the place where the grapes are grown. If it was just about the plant and winemaking techniques, the logic follows that you could take vines from a Grand Cru vineyard and plant them in a flower pot in Manhattan to make great wine."

If we have serious and contentious winemakers who make their wines all about the same way, how can the wines be so different from one parcel to the next? The magic is the place and the place possess the magic—they are inseparable. After all these years, I still discover little differences in the placement of vineyards and their subsoils that make the magic come alive. Are these differences by chance, by nature, or a bit of both? Consider with me: exposition, topography, heat, and weather.

I had recently read a book that I thought was going to be more practical, yes, romantic and not technical, called *Soils for Fine Wines* by Robert E. White. White is a soil scientist, so I was quickly in way over my head as he expounded on the ability of different chemical compounds to be absorbed or not depending on the soil, sun, precipitation, irrigation, and the timing of all with totally incomprehensible mathematical and chemical formulas. After slogging through 250 pages of his theory that terroir does not really exist, he makes the case in the last five to seven pages with a great many ifs. There is terroir, but there is a world of difference between wines with character and the mass of other wines made today that are of high quality but lack personality. He writes "with skilled wine-making techniques, it should be possible to make quality wine from disease-free, mature fruit of any desirable variety." (This is to say you can make good wine from anywhere as long as the grapes are healthy.) "But because soil is so variable in the landscape, it is likely that quantitative relationships between soil properties, yield, and the composition of grapes will only be elucidated on a local scale."

The idea that a growing area has specific characteristics that make it special should not be that hard to accept if we show examples such as Georgia peaches, Vidalia onions, or oysters from the Pacific Northwest as compared to those of the Chesapeake Bay or from Maine. We know they taste different, but should we not ask why? Just as there is a great deal of good food out there, there is a great deal of good wine, but this is not to say being a good peach or a good wine makes it unique. What we are saying with the word *terroir* is that there are special places in the world that under the correct conditions and with the proper husbandry produce unique flavors and sensations.

I intuitively knew this as I tasted and made Burgundy, but it was not until I had my own domaine and became focused on the farming that the reality of the dirt, place, and season became one. English has two words for vigneron but in French there are four. It is a curious word and is defined in *Lexivin*, the dictionary of wine as a: "vinegrower, winegrower, vintner, a profession." What is a winegrower? An accurate definition of a Burgundian vigneron brings to mind bottles literally growing on the vine, not the latest superstar winemaker, grape grower, or consultant that jets around the world as a high-paid hack. We are farmers and the process of making wine is synonymous with growing the grapes and husbanding our land. The two are inseparable and quite impossible to understand until you think of them as one.

Burgundy is the result of many small things, many tasks that build over a year and then speak through the glass. How we hear these sounds, interpret these tastes is individual but we all have to interpret them in our own fashion; that takes skill. Growing great grapes is not a romantic theory. It is viticulture, horticulture, and farming at its highest level. It takes great knowledge to husband grapes in an environment as dodgy as Burgundy. Climate/weather patterns have changed and we must adapt.

The Côte d'Or produces only one percent of the wine in France at the 47th parallel. What this means, in a simplistic way, is that at the height of summer Burgundy has 28 percent more daylight hours than Santa Barbara at the 34th parallel, but on the first day of winter, it has 28 percent less. In other words, Burgundy's growing season has a very steep curve, actually a steep chevron. Geographically as we get closer to the equator, the curve flattens. From a botanical standpoint, this means that the closer we get to the equator, the vines get more sun earlier, more sun later, and less in the middle of a growing season than in Burgundy.

In the mid-1990s, a college friend visited me in Beaune in the late winter/ early spring and asked, is the weather always this bad? Yep, I answered and this is the deep, dark secret of Europe. The French continental climate is due to the tempering effects of the Gulf Stream on the North Atlantic and this is why England has a mild climate and why Scotland, because of the North Sea, is cold and stormy. With this continental climate, Burgundy gets much more rain more evenly throughout the year than in many other wine-growing regions, including the Rhone in France, Spain, California, and Chile. We receive on average, two inches of rain per month. May, our wettest month, acts as our summer water reserve for an annual rainfall of approximately twenty-eight inches.

To understand the mystery of Burgundy's terroir, it is also necessary to know a bit about its geology. The rocks are the reason that we have the wine. Geographically, the Côte d'Or lies about two hundred miles southeast of Paris in east central France. We are actually closer to Geneva than Paris and we can often see Mount Blanc and the Alps at sunrise or sunset. Geologically, the Côte d'Or represents part of the western slope of a rift river valley separated by the Saône River and the Côtes du Jura, the foothills of the Alps. The Saône valley has been used as the interstate of France since Roman times: it runs north to south, so it allows for the easy flow of peoples, goods, and information from prehistoric times to the present. To its north, it connects with the Seine's tributaries; to the south, the Saône joins the Rhone at Lyon then flows to the Mediterranean.

The Saône valley is fertile with heavy soils from millennia of erosion from the opposite slopes that make the valley ideally suited for grazing and growing feed and grains. Our slope, the Côte d'Or (Golden Slope), an escarpment (actually, several) about thirty miles long with Dijon its northern terminus, is composed of two principal types of limestone from two distinct geological formations during the Jurassic period 195 to 135 million years ago: Bajocian and Bathonian (ancient sea lilies, dense and hard). These are often seen as Comblanchien limestone (marble) in the Côte de Nuits (northen half) and oolitic and Callovian limestone (coral structures) in the Côte de Beaune (southern half). During these periods, the environment was much warmer. Imagine what is now western Europe as a swampy shallow inland sea where the lilies and crustaceans lived, died, settled to the bottom of the sea, fossilized, and became a dense relatively heavy and non-friable limestone. As the continents and tectonic plates continued to move, the environment became a more tropical Caribbean-like environment, where the creatures were similar to coral. Both formations have a high level of calcium carbonate that react with seawater to create limestone. Simply put, there are myriad limestones in the Côte d'Or and it is their various arrangements—more, less, or not at all on the slopes—that gives Burgundy its complexity.

Terroir is not, strictly speaking, just the dirt or rocks, terroir, or a particular parcel's character, is made up of a variety of elements and these in combination with one another give a specific vineyard its character. To fully understand, we need to look at the French words that refer to a vineyard, its placement, and its personality and the nuances of this notion of terroir.

- Appellation comes from the French word appeler (to call someone or to call yourself, i.e., a name or its name). An appellation is the specific name or designation of a vineyard in Burgundy. It includes four vineyard classifications (levels) that increase in specificity (location) as you climb higher in classification: Bourgogne, Village, 1er Cru, and Grand Cru: or from 4th to 1st level, from a D to A grade, 60 to 100 points.
- Climat (pronounced "klee-ma") is not the weather but the combination of all of the factors that form a vineyard's terroir. Climat is interchangeable with appellation or again the name of the vineyard.
- Village or commune (not communist, but a community or the village) are the names of the tiny but famous villages you pass through in the Côte d'Or and within these designated geographical areas where the individual vineyards lie: the most important point of this entire discussion. Think big to small, general to specific (sometimes minuscule) and you will quickly understand Burgundy.

For instance, my domaine that in 2019 comprised thirty acres (twelve hectares) of vineyards was forty-five separate parcels. Again, this is typical in Burgundy and my domaine was no different than many other domaines. Visitors often asked to visit my chateau, always a laugh because most of our chateaux are our homes. To visit my vineyard, a "take me to your leader" question was another tough one. "Which one? I have forty-five," was my response.

At the risk of insulting some Burgundophiles, I will inundate you with metaphors. First, imagine the vineyards of the Côte d'Or as the profile of a sandwich with the designations of the vineyards and their rankings corresponding to the placement on the slope of the bread, cheese, and meat.

The 4th level, comprised of about 50 percent of the designated vineyards in Burgundy, is classified as Bourgogne, where you will find Bourgogne Pinot Noir (rouge) or Bourgogne Chardonnay (blanc) on the label. The Bourgogne vineyards are the lower or upper portions of the slope, the bread if you will. In general, they are the most difficult land to grow on, and on which to ripen grapes. The lower slopes tend to have soils that are quite heavy, humid with poor air flow, leading to more fungal disease, as well as higher farming costs with frost and hailstorms likely. The upper slopes have much thinner and meager soils, are much cooler and thus their growing season starts later in the spring and finish later in the fall as the days get shorter, with the threat of fall rains, and are therefore more difficult to ripen.

The 3rd level, the village classification, is about 30 percent of the designated vineyards. This is where you will find the name of the village or town on the label. With the cheese in our sandwich, we are getting closer to the mid-slope. In general, you will find this a double cheese sandwich because in most cases, the village appellations are higher up or lower down the slope than the Bourgogne-classified vineyards. The village vineyards begin to have more complex soils, better drainage, and aeration. The label will indicate the name of the town or village, where the grapes came from: Volnay, Meursault, Chambolle-Musigny, Beaune, etc.

As we climb the slope, we reach the 2nd level of vineyards, classified as 1er Cru, where you find the name of the village and the specific name of a vineyard. Here, we have the meat of the slope, composed of the most complex soils, best drainage, best airflow, and the most sun. The reason these sites are Premier Cru (and Grand Cru to follow) is because these vineyards, through trial and error for a thousand years since the monks of Citeaux began to grow grapes and make wine, have been found to consistently produce the most complex and complete wines especially in difficult vintages. Most importantly, and I cannot stress this enough, because Burgundy lies at the 47th parallel, our growing season is based on how many hours of sun we have in a growing season not temperature days used in more southern climes. The 1er Cru and Grand Cru vineyards sites have a better orientation/exposition to the summer sun, south/south-west, that allows them to capture more sunlight hours for photosynthesis. Managing the vineyard to maximize photosynthesis, turning the grapes' starches into sugars is the vigneron's raison d'être. Because we are at such a high latitude, about as far north as you can ripen grapes, each hour of sun directly affects ripening or lack thereof. Also, because we lose minutes of the sun's exposure at a rapid rate after the summer solstice, the mid-slope vineyards, the 1er Crus and Grand Crus get the day's first and last rays of the sun. This extra sun doled out gently over a growing season directly affects the quality of the grapes.

Finally, about two percent of the designated vineyards in Burgundy, only thirty-five vineyards total, are classified as 1st level or Grand Cru. With this classification, the name of the vineyard is exclusively on the label. This is the foie gras or truffle of Burgundy vineyards. What is clear is that the Grand Cru vineyards can and generally do produce, when farmed and made correctly, the wines of the greatest longevity, finesses, and complexity.

Now, think of Côte d'Or as a glass slide that you place under a microscope. As we place it under the instrument and begin to adjust the focus we see forms, cells arranged chaotically to the untrained eye. On further reflection, we begin to see that they are arranged north to south as small communities, villages but, most importantly, as distinct geographic areas. For example, in the south, on the left of the slide, the form of the village of Chassagne-Montrachet comes into focus.

As we increase the magnification, the boundaries between the cells are quite distinct, are carefully arranged, and are all as unique as snowflakes. Each's width and length is different and they even have different pitches and contours over the face of the slide; no two are alike.

If we increase the magnification and peer further into one cell, the village, we begin to see individual forms some large, many quite small that are its composite components. Here, individual divisions indicate 1er Cru and Grand Cru vineyards, such as La Maltroie whose land is shaped as an "L."

If we have an electron microscope and peer into the darkness of its DNA, La Maltroie's soil shows a geology of marls, limestone, and, most importantly, iron ore stones that give this vineyard's wine its longevity and steely structure. This does not alone explain why the wines of Burgundy are unique from any other pinot noir or chardonnay in the world.

Assuming that all the factors are equal, farming and winemaking (even though they never are), what distinguishes vineyards is almost an infinite combination of factors relating to soil, exposition, and the quality of the actual planted vines.

Finally, finding exquisite terroir or the most unique character of a particular vineyard is no different than the intrinsic elements and structure of music. There are great composers and compositions from Bach, Beethoven, Brahms, and Mozart. Their music has varying levels of complexity: dance, sonatas, chamber, opera, concertos, and symphonies. The notes exist, can be read on a page, and do not change, but what changes is who is playing the music and how. Just because I own the original score to a Bach cantata or own a

Stradivarius violin does not mean I know how to play the music or play the music any better than a high school band.

I now owned 1er Cru and Grand Cru land and I was determined to play the land like one of the greatest symphonies in the world.

Chapter 25

Les Nouvelles Mathématiques (The New Math)

Remember new math? Do the words bring foreboding thoughts to mind? Those of us over 60 relate. I never really understood the difference between new math and old math, except that the new math was hard. I actually looked up its origins and found out that in the end, it was a fad in response to the space race with the Russians.

New math aside, I did excel at basic math, not calculus, but the relationship between numbers that, for some silly reason, I can do in my head—percentages, profits, losses, and costs. This provided an advantage to me as I assessed the landscape from 10,000 feet. Even more so now than when I made wine, Burgundians are facing fundamental issues that are causing lower yields, higher costs, scarcer product, and at least a generation ahead of us to get back on track. If the vignerons of the early 1990s challenges were correcting the post-war mistakes of chemicals in the soils and producing better wine in the face of world competition, the vignerons of 2020s face a combination of production costs and climatic questions that are threatening their very existence. Despite the self-congratulations on Burgundy's success in the fifteen years from 2005, I started to reflect and talk to other vignerons about our rising costs, lower yields, and our inability to recoup these costs in the prices of our wines. I felt something was not right.

For example, 2009 was a normal year for production, but the next seven years were on average 27.5 percent less in volume than 2009, with 2012 down 41.6 percent and 2016 down 45.8 percent. A combination of factors relating to poor spring flowering (not unusual), warmer winters, and earlier springs due to changes in the climate plus periods of uncontrolled oidium and mildew (fungal diseases) pushed the cost of treating the vines through the roof. Throw in three years in a row of devasting hail, literally destroying whole crops, and we felt like Sisyphus, the trickster in Homer's *Iliad* who tried to cheat death.

I began to see that what some would call just bad luck was a combination of systemic and macro factors that had no short-term solutions and required a concerted generational effort to combat. The commitment to work on the macro pressures that were causing most producers, except for those blessed with 1er Cru and Grand Cru vineyards and a handful of domaines that could sell their wines at a premium, into an economic situation where unless harvest was close to full yields, production did not cover basic farming costs.

The consequences included:

- Very small harvests because of weather-related issues. We can place the blame on acts of God and his divine retribution but no matter what you believe, we must recognize that we did have a Biblical-like stretch of seven lean years and must plan, as business people who store grain (in this case, capital) to work through what I came to believe would be, at best, harvests of erratic yields and highly volatile farming costs. This is the new normal.
- Two times the normal rainfall in 2012 from March 20 to June 20, 50 percent less sun in June, and over the course of the season, treating 50 percent more vines for fungal diseases to try to salvage the few grapes that remained. This translated into the non-trivial burning of 40 to 50 percent more diesel fuel: so much for green farming.
- Green farming techniques; no chemicals, more plowing, etc., result in healthier soils, better grapes, better wine, lower yields, and higher costs, but can these lower yields be translated into more valuable products in the form of higher prices? Is the end consumer willing to pay up to 30 percent more for a green product? Today, the answer is no, unless Burgundy can restructure its distribution system with more direct sales to consumers to keep this added green value. Otherwise, will the traditional distribution system and market accept escalating prices that double our prices on a retail shelf? If not, we will be subsidizing the consumer's political and environmental proclivities with our labor and capital.
- Old vines must be replanted with more productive plants. Older plants generally do produce great grapes, but the quantity versus the cost of producing those few grapes is uneconomical (unless you have 1er or Grand Crus sites). If we assume a mortality rate for a vineyard under the best of circumstances at 1.5 to 2 percent per year, with three years for a new plant to reach production (if you are lucky), the results are at best a utility of just 94 to 95 percent of any given vineyard before any other climatic pressures. (It is scandalous that universities in France, the profession and industry are not focused on spending research dollars to develop new bio treatments and plants that are more resistant to viruses, diseases, and our changing weather patterns.)

- A rising demand for Burgundy all over the world. The rush for Burgundy has two speeds: the ten thousand uber rich people in every corner of the world who are bidding up the prices for the collectible Burgundies, and the rest of the 80 percent of the wines that are not 1er Cru or Grand Cru.

I wasn't afraid of getting my hands dirty to look at the real costs of farming a hectare but therein I found another paradox—why, even though the cost of the grapes is elastic, is our pricing inelastic. Hard as it is to imagine, the cost to farm a hectare of vines is the same whether it is a Bourgogne or a Grand Cru vineyard. It actually costs less for the 1er Crus and Grand Crus because of drainage, airflow, and the sun's exposure, leading to fewer maladies and expenses. If you truly account for overhead, all-in expenses are around 25,000 euros per year, per hectare. But in a year such as 2012, it can climb to 30,000 to 35,000 euros because of the extra treatments. With a fixed cost of 25,000 euros, it is easy math to see what we need to produce to meet our costs. This does not even consider rent and return on capital: just the pure direct costs. (These costs are based on 2018 numbers.)

Break it down further and here is the per-bottle cost of the grapes, considering farming costs of 25,000 euros per hectare (2.47 acres) at the following production levels (with one barrel equaling 25 cases or 300 bottles).

10 barrels: 2500 euros per barrel = $9.83 per bottle
15 barrels: 1666 euros per barrel = $6.55 per bottle
20 barrels: 1250 euros per barrel = $4.92 per bottle
25 barrels: 1000 euros per barrel = $3.93 per bottle
30 barrels: 833 euros per barrel = $3.28 per bottle

The logic of the math is simple: three times more grapes equates to cost three times less per bottle regardless of the appellation. What the above means in layman's terms is that the cost of producing our raw materials in a year where the yields are down by just a third can equate to 38 percent to as much as 58 percent of the selling price. Why can't we simply raise our prices? Because of the perception and price ceiling that is created by the appellation system. I sold many of my lower appellation, less prestigious wines, at the absolute limit of what the market would bear. In other words, to have a correct cost of goods at a maximum of 20 percent, I need to sell my Bourgogne Chardonnay to the importers for $30.00, not $11.00.

With the economics of production getting more and more difficult, I had to make a decision on how I would manage my six hectares (fifteen acres): continue to limp along and hope the good years would average out the bad years, buy more grapes from third parties if they could be found and tie up more capital, or grow my domaine to a size where there are some economies of scale in production. What became apparent was that just a few years before, six hectares (an average-sized, manageable, and slightly profitable domaine of 14.8 acres) did not work: I had to grow or die. I needed to double my acreage to make it function on a cost basis and to continue my goal of weaning myself off of bought grapes. Having more and more of my own production would allow me in essence to buy grapes from myself and over a three-year period, all the cash that was being sent to outside vendors would be kept in house.

I was provided the opportunity to further expand in 2012 as the year sent many a winemaker over the edge. In early 2013, my general manager told me that her neighbor, a vigneron in St. Romain, Christophe Buisson, was burned out from the extremes of the 2012 vintage and was ready to "péter un câble" (pronounced "pet-A," not "Peter"), translated: lose it. Christophe, who started making wine in 1999, set up shop next to my original location in Beaune. Over the years, he had become a close friend and ever present in our lives. We bought and shared our bottling line together and we gave each other a hand when in a pinch. An Energizer Bunny of a man, Christophe is 5′6" and races motocross bikes and his road bike for 75 miles at a time. He is personable, always has a joke, and is friends with winemakers up and down the Côte. He is also a real bavard, this is to say talkative. Always one to dramatize a situation and the repository of anything you ever need to know in Burgundy, I learned that there are no secrets with "Papa" as we affectionately call each other.

Christophe, who farmed his fifteen acres organically, met head on the reality of not having enough acres, nor the disposition to be able to hire a team to help him in the vines. In 2012, the cycles of anti-fungal treatments reached a cadence of once every four to five days, and caused us to run around like chickens with our heads off, chasing parcels scattered over fifteen miles north–south and two miles east–west—all the while having to manage the paperwork, sales, and collections. It is amazing more vignerons did not péter un câble. I remember as though it was yesterday, Christophe running into my office grabbing an espresso and repeating "C'est la merde, c'est la merde, cette saison est un bordel, et en plus mon tracteur est en panne et il n'y a pas des pieces disponibles!" (It's crap, it's crap, this season is a bordel, and now my tractor has broken down and there are no parts!)

With a hail storm at the end of June that wiped out 100 of the year's crop in multiple vineyards, the attitude at harvest was simply, let's get this over with and move on. Somehow, we survived 2012, but the scars and exhaustion affected everyone and were not far from our minds in 2013 and 2014 when two distinct hail storms, both in July, cut a swath through some of the same vineyards as they did in 2012. Hail stones the size of moth balls literally denuded the vines of leaves and grapes in less than a minute. Leaves were shredded and the grapes torn off or severely damaged, leading to rot, and bruised and damaged vines and shoots for the following year.

With this as perspective, I spoke to Christophe in July about what I had heard and asked him point blank "Christophe, qu'est ce que tu voudrais faire? Est-ce que le métier te plaît toujours?" (Christophe, what do you want to do? Do you still like making wine?)

"Je suis cuit, il faut que je change ma vie. Il n'est pas possible de continuer comme ca. Les vacances commencement bientot, on parle après la vendange." (I am fried, I have to change my life, I cannot continue like this. Vacation is starting soon, let's speak after harvest.)

That fall, we continued our conversation. In December, I presented him with an offer to buy his vineyards, half of which he owned, primarily in St. Romain, which is known for its whites and his third-party leases for the other half. With winter underway, this was the perfect time for a sale like this. Christophe and I agreed quickly on the outline of a transaction where I would buy his vines and he and I would become partners in the leases he controlled beginning in 2015. This transfer was a great deal more complicated, time consuming, and involved than I am explaining, but can be summarized in one word: trust. Christophe trusted that I would perform what I promised and the landlords trusted that Christophe would continue the rent payments albeit from a new joint structure. Throughout 2014, we defined fifteen different leases, their conditions (several needed replanting), examined lease terms, valuing them, and secured landlord approval.

My domaine was a collection of land owned through land partnerships of which I was a significant partner and also through third-party leases. What is often misunderstood about Burgundy is that a domaine is defined as the land/vineyards it directly farms. It includes land you might directly own, land via intrafamily and interfamily leases, land held in usufruit (trust), and third-party leases. The French term *usufruit* literally means user of the fruit. We would say that the holder has a life estate and receives the income from the vines as long as he/she is alive. At death, the right to the income reverts to the fee holder. This is a popular way of handing down vine ownership because it splits out the value of the land between its income and the fee interest and reduces inheritance taxes.

It is difficult if not sometimes impossible to follow the holdings of many family domaines that seem to be created or transformed with every new generation. Just take for example the village of Vosne-Romanée, where many domaines have morphed for a variety of reasons. Domaine Jean Gros, originally one very large domaine, over the last generation has become four, wonderful, high-quality domaines divided among cousins who are now all in their late 50s to mid-60s. What will happen with their successions? Louis-Michel Liger of Domaine du Compte Liger-Belair, who trained not as a vigneron but as an engineer, returned to the family domaine as its leases and grape contracts were expiring. He brought all their operations back in house to spectacular results. Finally, there is Domaine Mugneret-Gibourg. Because of Dr. George Mugneret's profession, he leased out most of his vines and made small amounts of wine. After his untimely death in 1988, his daughters began to run the domaine with extraordinary results as they began to make wine from their grapes that were often under sharecropping contracts and took back land as the leases terminated. In 2016, a vigneron to whom his father had leased the land decided to retire and turn four hectares back to them fifty years after the initial lease. These are not isolated cases but are Burgundy's landscape. It is an ongoing process as new blood arrives, older vignerons retire, or, as can be the case, new blood arrives to disappointing results. These are the normal lives of any family enterprise, so it is not surprising that I acquired many of my vines because of generational changes.

Now that Christophe, one of the new generation was burned out, my work was to keep him focused and on message. Did I mention that Christophe is a constant worrier and my mantra with him was, "Ça va aller, apres le premier paiement des locations les propretiers n'y penseront plus" (It is all going to work out, after the first rent payments, the landlords will not even think about it again). We closed on the transaction in March 2015. It included me buying his three tractors, plows, and a sundry vineyard equipment and garage for storage. As we shook hands and signed the papers, a grand smile and look of relief appeared on Christophe's face. I too smiled, knowing that I had in one acquisition doubled the size of my domaine and achieved my goal of the last ten years to become a domaine that should on average produce 300 barrels of wine (around 7,500 cases per year). I was no longer dependent on the whims and pricing of third-party suppliers.

As I integrated Christophe's vines into my domaine, harvest 2015 surprised me with another equation—a production year that was average in size but was of extraordinary quality. However, no amount of planning, praying, or lamentations could prepare me for 2016.

Chapter 26

Un Matin Noir en Avril
(A Black Morning in April)

In June 2016, Elin McCoy the wine writer who also worked for *Bloomberg News* asked me what was happening in Burgundy after an unprecedented April frost. "Imagine if you told Mike Bloomberg that in three of the next six years you will have no income and have the same costs because of weather like this. What would his reaction be? At minimum, he would be apoplectic, say you are crazy, or might even run again for mayor of New York."

As my business grew, I recruited and trained a new generation to do the hard work. I was there, watching everything, and still operating the forklift during harvest, which made my team very happy because "le boss" was staying busy and not messing with the team's minute-to-minute harvest execution. This new arrangement pleased me to no end. I no longer had to be everywhere all at once but could step in when needed.

In Burgundy, there is no seminal moment when you show up to turn water into wine; there is a twelve-month process that ends at harvest and begins again after harvest. Harvest and the fermentations are a six-week period, a bit like the NFL playoffs. The rest of the work is a continuous cycle that pre-pares you to execute and adapt to the vintage. It is not linear but it is cyclical, punctuated with stops and starts. Each harvest and growing season are inter-connected; there is cause and effect. This is why the ability to adapt—only learned through experience—is what creates either average or great wine. Notice I use the word *harvest*, not *vintage*. There are many different choices to make at each harvest and a winemaker's ability to make sense of what is going on is the challenge and the excitement of harvest.

The concept that one year is connected to the next is best visualized as a continuous spiral. In concrete terms, what happens in 2012 will have effects in 2013. For example, the formations of the year's buds (embryonic grapes) are actually formed in the prior year and their fertility, or lack thereof, is a direct result of the previous spring's weather. If the spring is cold and wet, next year's buds are normally fewer in number and less fertile.

In 2016, weekend weather forecasters announced cold to freezing temperatures for the next week beginning Monday morning April 25. Friends in London reported light snow and freezing temperatures on Saturday night and the weather patterns showed the cold front descending toward France. Imagine a relaxing Saturday evening watching the news to hear "By the way, a tsunami will hit your home between dawn and 10:00 am Wednesday and there is nothing you can do about it."

We had a lovely, warm and dry April and the vines were well past their third or fourth leaves (growth) especially the chardonnay, with two to three leaves for the pinot noir. Most critically, the infantile grape clusters had begun to form and were exposed to the first wave of below-freezing temperatures that descended and damaged the Savigny-Chorey-Pernand-Ladoix vineyards, just to the north of Beaune, Monday morning. That afternoon, the weather was lovely, clear and crisp, more March like than April, but after Tuesday morning's cold temperatures and scattered damage, the humidity rose and we expected precipitation. My college roommate, his wife, and friends were visiting, so we invited them for dinner. As we drove to my home in Orches at about 7:00 pm, a light, wet snow began as the temperature dropped to 32 degrees. Not good, damage to be expected, but manageable. When they left at 11:00 pm, the snow had stopped, the skies cleared, and the temperature continued to drop. This was the beginning of the disaster as the early morning temperatures dipped to 28 to 23 Fahrenheit (−2 to 5 degrees Celsius).

Black frost formed as the sun rose because the moisture in the early morning hours froze on the leaves and buds. This would have injured the vines, but the crucial factor was that the frost was followed by clear blue skies. The rising sun through the prism of the thin frost layer literally burned the young leaves and buds. Like the science experiment when you focus a magnifying glass on a leaf in the sun or the accidental time you put lettuce or basil from the sunny farmers market in the back of your refrigerator and it got too cold, burned and then turned mushy.

If the skies had been overcast, which they often are in Burgundy, the temperatures would have fallen less and risen more slowly and melted the frost with the cloud cover protecting the vines from the burning rays of the sun. It is counterintuitive but when orange farmers in Florida spray water on their trees to protect them from frost, they literally are forming tiny protective igloos on the nascent orange buds. This ice cocoon protects the bud from the outside cold and from the sudden burning effects of the sun's rays. Water freezes at 32 degrees (0 degrees celsius) and does not get colder. Thus, the ice coating acts as insulation from the outside cold and as the temperature rises the protective cocoon melts away. This is not the case with black frost, where the ice coating is so thin that it is devastating when partnered with the sun.

I did not go out to into the vines Wednesday. There was nothing to do, but I looked at them on Thursday and already saw that the leaves were a ghostly white with black tips and deformed buds. We, the royal we, gamely and optimistically said on verra (let's wait and see). It takes time to see the full results, but in our hearts, we knew. After two weeks, it was clear we were in trouble. A month later, the full extent of the catastrophe was evident: no buds, no leaves, and necrosis on the canes and the infantile shoots.

The frost was unprecedented in its reach and duration; it caused far greater damage than the last large frost in 1981, with some older vignerons saying the frost was more reminiscent of 1956 than 1981. It is estimated that 80 to 90 percent of the Côte d'Or's surface was affected. Chablis and Champagne often get frost but deal with this regular climatic condition by pruning longer and leaving more buds. This increases their odds of surviving frost by leaving more future grapes. Their best vineyards also have sprinkler systems and smudge pots-to battle cold temperatures. When Burgundy gets frost, it tends to be a one off in a specific area and even then, it is rare. For most people in the Côte d'Or, this was hopefully a once-in-a-lifetime experience.

After the initial shock, we returned to the vines and turned the soil to encourage a second bud growth. There can be a second set of buds, but there is no rule of thumb as to their quantity or fertility. The vines were also in shock and after a month to six weeks, parcels resumed their normal growth while many were just beginning to show minor recovery. The frost created even more work in the vines as we tried to salvage and husband the buds and shoots that remained for the year and train them for next year. This is a tedious and highly skilled job that requires the vineyard hand to examine each vine as an individual patient, determine where to cut off unproductive shoots and suckers, and decide which ones to leave for next year's vine. This coupled with unprecedented mildew pressures created skyrocketing treatment costs for significantly fewer grapes. When asked by an American friend what this meant, I said: "I lost a bit over 50 percent of my grapes and it is going to cost me $400,000 to 500,000 dollars. In gross bottle sales, it cost $800,000 to 1,000,000.

"Are you serious? Come on, it cannot be that bad," he replied.

"I grow around $600,000 to $700,000 worth of grapes a year, and I lost 50 percent, but what's most painful is that I lost 80 to 90 percent of my most valuable grapes." (My vineyards in Bâtard-Montrachet, Puligny-Montrachet, and Chassagne-Montrachet.) Vineyard treatments for oidium and mildew were so exorbitant that several well-known bio producers threw in the towel. We stuck with our bio regimen because unless we switched from conventional spraying early in the season, the efficacy of the switch was marginal or non-existent.

These dark frost days were the opposite of the sun-drenched days of sipping wine that my friends imagined I had as a vigneron. Only my great friend and business partner Andrzej Rojek, who emigrated from Poland just before the rise of Solidarity and still has a wonderfully thick Polish accent, understood. When I saw him in Brooklyn in February 2017, he somehow expected me to change my story of 2016 as well as the hail damage of 2012, 2013, 2014, and small yields in 2010, 2011, and 2015.

"Alex, stop it, you are making me depressed and making me want to kill myself!"

Percentages, statistics, and averages never do justice to the real pain that is inflicted in cases like these because they mask the extremes. After the harvest, I spoke with many colleagues along with reading the official statistics from the growers' association. The losses ranged from 90 percent to none at all, thus the tyranny of averages.

The numbers were depressing, especially in the context of the year and most importantly how it completed our seventh lean year. Production for greater Burgundy, from Chablis to the Mâconnais, was down 20 percent, with Chablis down 47 percent and some winemakers not making wine from many of their best vineyards. Côte de Nuits' losses were 18 percent, but that masks a 53 percent loss in Marsannay and a 39 percent loss in Chambolle-Musigny, not to mention significant drops in Gevrey-Chambertin and Clos Vougeot and Gevrey-Chambertin, where we made 50 percent of what we normally produce. We were only able to pick a few boxes of grapes in the 1er Cru Les Charmes and Chambolle village, from which we could not make any wine.

The Côte de Beaune was down 24 percent in production, with Volnay village down 60 percent, and St. Romain Blanc down just 16 percent but 50 percent in its lower-elevation vineyards. Behind the numbers, like the six-foot-tall man who drowned crossing the stream that was five feet deep on average, there were crazy random occurrences that were hard to process. While Volnay village wines lost at least 60 percent, Volnay's 1er Cru parcels were spared and produced a normal full crop. Meanwhile, the same quadrant on Volnay's lowest slopes, my Pinot Noir vines (Bourgogne) lost everything. In fact, after the frost, I told my team not to spend another moment in our best Bourgogne parcel except to pull out the vines in the fall and replant because production was already close to nil after three consecutive years of hail and a fourth of frost. Subsequently, many of our neighbors did the same in other low-lying vineyards. In the two years following the 2016 frost, I had never before seen so many vignerons in in the process of replanting vineyards throughout the Côte d'Or. This renewal process takes five to seven years to get back to full production levels.

Growing grapes and making wine requires a commitment to the long term, of living within and often under ones means, getting dividends from unlikely sources, and always being a farmer. It is not an easy place to live and work. The French say it requires des reins d'acier (steel kidneys).

Unimaginable as it seems, in 2017, after a warm and precocious March and April, frost threatened us again. The week before, virtually all of the local grower associations organized a decidedly low-tech battle plan. They bought large Ho Ho–like hay bales and placed them strategically in the vineyards about fifty meters apart with everyone volunteering to burn four to six bales if the temperature dropped below freezing. The goal, creating a man-made cloud of smoke throughout the Côte d'Or was achieved on Saturday April 29 when, at 5:00 am, we received text messages to man our positions, drove to the vines under cloudless clear skies and temperatures ranging from –2 to 3 Celsius.

Within forty-five minutes, the sky was blocked by our smoke screen. The sun rose at 6:45 am and helicopters fanned the vines to circulate the warmer higher air. The temperature rose to 2 degrees Celsius (35 to 36 degrees Fahrenheit) under the smoke by around 7:45 am, saving the crop. By the way, I strongly recommend wearing a surgical facemask if you are ever presented with such a challenge. Fighting or creating fires is its own profession and not one of my normal job requirements. I was coughing for two weeks afterward.

PART 7

La Vendange

Harvest

The visible results of a year's work are revealed in a compressed four-week period. Vineyard ripeness is monitored, harvest and winemaking equipment is cleaned, new barrels are received, old barrels are cleaned, and harvest picking teams are assembled. Think of harvest as a well-oiled machine that sits idle for eleven months but comes to life with aplomb and instinctively adjusts its speed to the harvest's cadence.

Chapter 27

Le Jeu a Changé
(The Game Changed)

The frost of 2016 just about put me over the edge. My plan of transforming myself into an independent vigneron was accomplished on paper, but financially I was just beginning the process. My ultimate goal was product and financial independence in three to four years of harvests. After that, I would be producing my grapes for free: the cash I would spend to buy my grapes would be my own. With good vineyard management, I should have a positive cash flow assuming stable sales (a dangerous assumption, I admit) and a turnover of physical stocks every two-and-a-half-to-three years (eighteen to twenty-four months of a new vintage in barrel, just bottled wine, and a current vintage for sale). This three-vintage cycle, which was already precarious after five small harvests, collapsed with the small harvests of 2015 (–30 percent) and 2016 (–55 percent). Simply put, this meant that my business plan would need at least another full year of cash to bring it to completion: cash either from my pocket, investors, or the banks because the wine that was supposed to become cash did not exist.

Beaune seems to literally roll up the sidewalks after the Hospices and we immediately start thinking about Christmas. Not only does Burgundy get quiet but the fog makes us dream of getting away. Our Christmas escape to ski in the Alps morphed into us returning to the States to visit family. In 2013, we found we could have both in Jackson Hole, Wyoming, so we made the Tetons our new home in the States.

After Christmas 2016, the New Year brought several sales trips that I was glad to hand off to the next generation. I joked that although clients still wanted to see my face and shake my hand, I was getting bored. Perhaps it was the brouillard (fog).

Burgundy's deep dark secret is its cold, damp fog from November until March. Because of our location to the far east of France's ocean climate (and winds) and west of the Alps, a large part of eastern France gets trapped in inversions (low-pressure weather systems that trap colder temperatures below the clouds). With the Saône river running virtually north–south, providing an added layer of moisture, cold, damp fog rolls in and often will stretch all the way up to my home since 2008 in Orches.

The beauty of Orches is that at 1650 feet (515 meters), we are often at or just above the line of fog that stretches from Beaune and burns off slowly. It is oriented almost dead south (as are so many of the best vineyards), so we actually get a great deal more sun than anywhere else in Burgundy. Tucked into the escarpments that ring the western border of the Côte d'Or, it is one of the most picturesque villages in Burgundy. When we are above the cloud, it appears like the heavens are below and the hills above St. Romain are islands. But we are not immune to the dampness.

"I do not want to spend another damp, cold, grey winter in Burgundy. We take a vacation every winter and go on two sales trips, let's do it stateside and not have to travel back and forth across the pond," I said to Diana. I was not talking about retirement, but the wear and tear of travel. "I have put a very good team in place—winemaking, vines, and administration—they do not need to see me every day."

I had worked myself out of a job and was ready to hand off my sales trips but had another year yet to navigate to reach my goal.

Before my return to France in January, I had a trip to California, then a Washington, DC, junket with all the associated delays and nightmares of winter travel through Chicago and the Northeast. These trips were also intended to introduce clients to Alexandre Brault, my COO (a fancy title for a small company to be sure). Since joining me in mid 2014, Alexandre was more and more in charge of sales, administration, and finance. As my banker, we had worked on several financing packages together, but it was clear to me that wine was his passion, not banking. I was pleased to have Alexandre begin to manage day-to-day administration and take over sales. At nearly 60, I was more than done with pounding the pavement and could not stomach any more nightmarish sales trips. Besides, many of the buyers were Alexandre's age, so I decided to show him the ropes.

In February, Diana and I took a dream of a trip to Sudan and its ancient Egyptian culture (I am crazy about Egyptology). After a stop at home in Burgundy, I began the whirlwind road show with Alexandre, starting in Florida, the US's second-largest wine market. For twenty years, I spent a week in Florida to visit my father and to participate in another over-the-top wine event: The South Beach Food and Wine Festival. My distributor, Southern Glazer's Wines & Spirits, founded the event as a small gathering to support the local hospitality college and was delighted when it grew (morphed) into an extravagant retail trade show, sponsored by the Food Network and *The Wine Spectator* magazine. There were multiple venues, tastings, dinners, and after parties, capped by a gigantic tasting on South Beach under the biggest tent I have ever seen. Guests paid big money to eat and drink non-stop and dress in their Sunday finest or not much at all. The highlight tasting was in the Fontainebleau's ballroom and *The Wine Spectator*'s "The Best of the Best" tasting. At the Best of the Best, Diana and I saw more stiletto heels, low-cut dresses, short skirts, long legs, and enough plastic-surgery enhancements to fill a gigantic ballroom.

I would often conduct tasting classes to engage people in Burgundy's magic and glue customers to my brand. I had good slides of vines and rocks and sometimes brought in another winemaker, my favorite being the late Jim Clendenen from Au Bon Climat winery in Santa Barbara. Jim, a Burgundy freak and a great American pinot noir and chardonnay winemaker, had long, blond, curly hair that swirled in all directions even though was in his late 60s. He was a force of nature with or without a microphone: think of casting Robert Plant as a winemaker. After one tasting, where all two hundred people in the audience were howling with laughter, Jim and I became a hit duo. Not quite Tina Fey and Amy Poehler, but more akin to the old Saturday Night Live Weekend Update with Dan Aykroyd and Jane Curtain mold of "Jane, you ignorant. . ." Our repertoire was natural as we entertained as much as we educated—the definition of good sales. We were asked, "Do you do seminars often together?" Being two dumb guys, we answered, "No, this is the first time." If we had been smart, we would have booked ourselves on a worldwide wine comic tour.

Alexandre and I, always on a budget, could not find an affordable hotel room in Miami Beach (you can never sell enough wine at 15 euros a bottle to support a $500 a night hotel room). Searching from Beaune, Alexandre found the Seagull Hotel, just off the beach in what seemed to be a good location. However, when we arrived, we found our hotel's entrance was adjacent to a public parking lot with the two police observation trailers. In defense of his decision, the picture on the internet did not show the lot nor the trailers (the photo Alexandre showed me was taken from the lot and made the hotel look 1960s Rat Pack cool). I figured at least it would be clean—silly me. To say we were staying in fleabag is an exaggeration: elevators broke intermittently, blood was on the broken tiles at the hotel entrance, the beds had what appeared to be original sheets from the 1960s, and windows screwed shut from the inside on the eighth floor. The next morning, I woke up early after sleeping on top of the sheets, I went for a long walk and a dip in the ocean for a salt water cure. As my American friends say, "such is the glamorous life of a French winemaker." (Not long after our sejour, the Seagull closed and since has been renovated.)

New York followed Florida with another trip across the pond in the second week in March for La Paulée (harvest party) of New York. La Paulée in

Burgundy is the grand finale fête of the Hospices weekend the Monday after the auction in Meursault. The world's biggest and best BYOB party with the best vignerons in Burgundy, headed by those from Meursault. It is a six-hour party for seven hundred guests, where the bottles are literally being passed at such a furious pace that your head spins. It is not uncommon to taste over seventy wines in an afternoon, assuming you are spitting and not just drinking.

New York's first La Paulée was created by the sommelier Daniel Johnnes in 2000 to re-create and spread the spirit of Burgundy and her wines to a wider audience. I was invited to his first Paulée and it grew into a major business for Daniel. By 2017, the event was decidedly more highbrow with tastings in the Rainbow Room and the final dinner at Pier 60 on the Hudson River. It was all first class and great fun to share a little bit of Burgundy with my New York friends. I must admit, however, it was a bit irksome to be stuck in the cheap seats in the back of the room at the final dinner. After twenty years of making Burgundy, I thought to myself, "Am I chopped liver?"

At this point, trying to figure out what time zone I was in was a moot point because I still had a trip to Oregon and California before returning home on March 21st. The day I returned, I received a phone call from my buddy from Meursault Sylvain Dussort, the Sylvain of the wine geyser barrels. Sylvain told me that a friend of his, who I had met at dinner at his house, was thinking of retiring, selling his domaine, and that I should give him a call. His domaine, Charles Allexant, was not at all on my radar but as I dug deeper, I was shocked by his holdings and how perfect they fit with the wines I was already making. There were vineyard holdings in the Côte de Nuits, Vosne-Romanée, Chambolle-Musigny, and Gevrey-Chambertin, normally unobtainable that were exactly the same wines I was making. In the Côte de Beaune, several reds and whites including a Beaune 1er Cru les Grèves and our carte de visite (lead white wine), a chardonnay from the area of St. Aubin, the 1er Cru Les Murgers des Dents de Chien. Dents de Chien literally translates as a vineyard full of big rocks shaped like dog's teeth and is still one of my favorite terroirs. I made multiple phone calls to set up a meeting, but he continued to dodge me. Little did he know that once on the scent of a deal, I was as tenacious as a chien du truffe (truffle dog). I was not to be denied.

After many friendly calls talking to his wife, about kids, the weather, sales, our dinner at the Dussorts, and even some drop-ins to get my face glued to my name, Patrice Allexant, his wife and I met to discuss their operation. They talked about the need for the transaction to be tax efficient and their desire to keep a hand in the operations. But I knew it would be simply a show-me-the-money moment. Not that I think that is wrong, I would have done the same thing. At the end of the day, business is business, and no matter what country you are in, it is always the sales price that counts, especially when it is more than your neighbor could get.

The key to getting a real understanding of any situation in Burgundy is to either break bread with or to taste the wines with the person in charge. It is amazing what you can learn by tasting someone's products. The tour of the factory is no different than due diligence in any profession: are the tractors in good condition, is the office organized, is the winery clean or are there strange smells, are the tanks new, has there been investment in new materials etc.? This does not tell you if the business is about to collapse, but it does give you a sense of the pride and respect the owner has in his or her operation. The Allexants easily passed this first test, they had a well-run operation, were on top of the details but were burned out because they did everything.

I asked Patrice, "Qui fait tes traitements?" (Who does your vineyard treatments?)

"Moi."

"Que vous?" (Just you?)

"Oui, je commence a 5:30 et finir vers 19:00 si le temps est bon." (I start at 5:30 am and finish around 7:00 pm if the weather is good.)

I thought, "This guy is tough and how in the hell how has he not blown a gasket?" As we tasted through his wines, I could tell they were clean and correctly made, but lacking emotion or les larmes (tears). The wines were not of another dimension but simply correct, which in Burgundy means a passing grade, nothing more. I glanced at his price list and thought that these were very inexpensive wines.

"À qui vendez-vous vos vins, quels sont vos marchés?" (Who do you sell you wines to, what are your markets?)

"Nos ventes sont 90 percent aux particuliers, la plupart aux foires de vins après les vendanges jusqu'à la fin d'année." (Our sales are 90 percent to individuals, mostly at wine fairs from harvest until the end of the year.)

"Mon dieu," I thought. They are on the road nearly three months doing the Namur tour! God love'em.

This last piece completed the puzzle for me as I saw that his domaine fit exactly into the mold of many of Burgundy's unknown domaines: they possessed enviable parcels but their products were simply good, not spectacular. The vineyards had been acquired over the years, first by his father, Charles, and then by himself—a typical story. His location to the east of Beaune, in the plain, for any customer looking to visit and taste, "au milieu de nulle part" (in the middle of nowhere), meant they had to go on the road selling. Finally, his domaine was one of many that could have moved up to the next level of quality with a few tweaks and adjustments, but only if their children were interested. This is the reason I had been able to acquire much of my holdings: there was no interested successor.

We spoke in detail in early April and I was vaguely optimistic I would be able to offer an attractive price for all the assets (this was another one of those deals that included the vines, equipment, building, stocks of wine, and payables and receivables—a mess to sift through to extract the jewels) and negotiate an agreement before a feeding frenzy began. I kept up contact, but it was always, "I am working on a plan with my accountant." In mid-June, I had pretty much forgotten the whole transaction when his new conseiller (advisor) called me.

"Merde," I thought, his advisor was a recently disgraced large négociant who was forced to sell his business because of fraud and had transformed himself into a land expert and sales guru. Within a week, we met at our offices and I immediately knew we were cooked as he droned on.

"Vous savez que son domaine est un bijoux de la côte, les vins et les vignes sont magnifiques, les bâtiments sont parfaits pour le tourisme de vigne les prix doivent le refléter ca. Donc à mon avis le prix dans une fourchette de 17 to 18 million euros est correct." (You know that his domaine is a jewel of the côte, the wines and vines are magnificent, and his buildings are perfect for wine tourism. The price should reflect it. I believe that a range of 17 to 18 million euros is correct.)

Beyond the fact that his numbers were bullshit, he was really saying: "You better open your wallet up because I am going to take more than my fair share from him and get ready for my Dutch auction." My back of the envelope calculations told me an 8–9 million euro price would be aggressive based on what I had bought my vines for and other recent market sales. After Alexandre and I put a sharpened pencil to the package, I said: "We could stretch to 10 million euros, but honestly I am closing my eyes and hoping for the best at that number. Alexandre, you know that we really have zero chance of competing and that I really do not want to waste my time with this jerk of a broker."

"Let me think about this, there might be a way to make a deal," Alexandre said.

"Have at it," I replied and to Alexandre's credit, over the next month, he put together a unique proposal.

The issue with vineyard transactions: is how do you separate the good from the bad? To put it bluntly, where can you make the most money with the least cost and hassle? The French term *valoriser* (to enhance or create value), is a more elegant and precise sense of the question. In any portfolio of vines, there are vineyards and wines that become the calling card of the domaine. Then there are other wines that you love to make, love to drink, but that take forever to sell: wines that your customers just do not get. This second category of making what the artist wants versus making what your market wants is every great vigneron's Achilles heel. How do you square this conundrum? You do it by creating a buying team that creates strength over the whole vineyard portfolio so that you do not have to discount the vines to reflect the wines that you will struggle to sell. Instead, you can put a proposal forth that creates strength across the whole portfolio.

Alexandre structured our offer based on splitting up the vines and capital needed with two other great vignerons that played to each's strengths:

- David Croix and Domaine de la Croix: David (in his early 40s) had been a friend and colleague for close to fifteen years. Originally from the Loire, tall, bearded, bespectacled, with a droll smile, he had always bought and traded grapes with me with a sense of fair play. In 2005, he established his own Domaine and in a short time became one of the references for Beaune's and Corton's vineyards. David's focus would be the Beaunes and some of the Côte de Nuits.
- Jean-Marc Roulot and Domaine Roulot: Jean-Marc, tall, initially taciturn but chatty and quick with a joke once he knows you, returned to the family domaine in 1989 after pursuing an acting career. His father, Guy, built up one of the iconic domaines in Meursault with multiple village and 1er Cru holdings that Jean-Marc, in turn, made into some of the greatest references of White Burgundy. He continues to act, appearing often in television and movie dramas and we became friends in 1999 as he was separating from his wife, Alix de Monthille. This was the period when Alix was working for me. Jean-Marc's focus would be the Puligny-Montrachets.
- Finally, I would focus on the vineyards in the Côte de Nuits and the St. Aubin 1er Cru Les Murgers des Dents de Chien (Teeth of the Dog) in the Côte de Beaune.

With this basic structure in place, we could each raise respective capital for a much smaller numerator, knowing that our investors were already predisposed to each's vineyards choices. We would create more value and could present a more aggressive offer together than apart.

The final offers were due before the August holiday and our package totaled 11 million euros. With a creative proposal, solid partners, and financing, I felt we were competitive in our offer knowing the feeding frenzy that was becoming Burgundy's vineyards. I could feel, hear the echoes, and see the money that Burgundy was attracting. More and more high-end tourism was transforming the economy from a sleepy local and regional affair into a global year-round chase for the best and rarest wines. The transformation was fast and the rate of change was even faster. Insatiable demand for ten percent of the total production as reflected in the feeding frenzy for the most prestigious vines and a "The Little Engine That Could" for the rest.

We heard back almost immediately that our offer was too low and would we like to submit a new one? We had made a very aggressive offer and that there was no way I was going to enter into a Dutch Auction. My partners agreed with me and we said, "Nous avons fait une offre plûtot aggressive mais juste. Il faudra voir comment les choses achèvent." (We made an aggressive and fair offer. We will have to see where things end up.) We quickly heard through several back channels (there are no secrets in Burgundy) that the top offer was more than 13 million euros (at least 20 percent more than ours and 30 percent more than my stretch offer). The successful bidder was new to Burgundy and was in the process of building a new domaine in Meursault: an heiress to one of the French supermarket chains, with virtually unlimited funds. We had lost before we had ever begun to compete.

As I reflected on the transaction, I became despondent about the opportunity to continue to grow my business.

- Did I have spare reserves of 10 million euros at a pop to buy a domaine: no.
- Did I have an angel investor who wanted to place 10 million euros every few years to buy vines: no.
- Did I have a group of very successful and supportive investors who might pony up 200,000 to 300,000 euros for an addition to their vine investments but not more: yes.
- Did I want to go on the road, hat in hand, finding new investors to raise 10 million euros: no.

Over the years, I developed a method in my decision-making process. To make a major decision, I learned to say no to many opportunities. By saying no enough times, simply through the process of elimination, great opportunities have appeared. When they do appear, I do not hesitate and yes comes with a sense of purpose. This was a similar moment of clarity but in reverse. It was clear to me that the game had changed and I could no longer compete; it was time to get out.

Chapter 28

Sinking (Affaisement)

My disenchantment with losing a bid for new vines was softened by 2017's full harvest. It lifted the spirits of all of us vignerons and our bankers. The large and qualitatively good harvest, especially for the whites, did not solve anyone's immediate cash-flow needs. But it did supply ample stocks in each winemaker's cellar so that the local banks felt comfortable in extending lines of credit for the coming year. This accompanied with brisk sales of the 2015s allowed me to breathe a momentary sigh of relief vis à vis cash flow: I did not have to come out of pocket to fund operations. This, was the third year with my team in place and I truly did play a supporting role, allowing mes gars (my boys), really my guys and girls, to run the show. I was there to make sure no one tripped up, but I felt content and proud of the quality work they were doing, for which to at least a small degree, I was responsible.

As the post-harvest calm subsided, we revved up for November and the Hospices feeding frenzy. The week seemed especially manic with so many dinner invitations that I had to wave a white flag. At a special dinner-and-wine event that we are invited to every few years, I literally hit the wall, but not from lack of calories or carbohydrates. The theater of the fall and the folly of the Hospices made me physically ill. I was so exhausted that I decided I did not want to do it any longer. As we returned to Jackson for the holidays, I looked back at my travel throughout the year and tallied the hordes of visitors who had come through, the tastings, the tours upon tours, and I realized it was no wonder I felt ill.

When I give a tasting, a seminar, or do a wine dinner, it is much more than show and tell. I want my guests to not only learn some of Burgundy's basics and have fun, but to come away with a bit of magic. I want them to drink the Kool-Aid and get hooked. If my guests glean an ounce of my passion for Burgundy, its wines, its nuances, its magic and mysteries, I sleep well. To me, it becomes a competition, but to do this requires a great deal of energy.

As I tasted wines with my guests, I often broke bread with them, initiating them into the way Burgundy should be enjoyed, which is over a meal. I wanted to show, not just tell people about the exclusive heights my colleagues and I wanted to achieve with our wines. I wanted them to feel the risk and the reward. In making wine, everything can go wrong. The joy in taking the risk to make something special is to faire plaisir (give pleasure). I think it's fair to say that this joy is something that all winemakers share. When I see a smile on a taster's face or receive a compliment like "This wine is taking me to a new place," I would give the wine away; this is recompense enough. The joy is the magic of making wine.

But alas it is a business. To grow the business would require me to double down on investing in new vines. At 60, I did not want to go, hat in hand, raising more funds. I was over the constant travel, more often than not in economy seats. Many of my clients, through some severe triage, had become dear friends, but I did not need nor desire to share another great bottle or great meal with them unless it was purely for pleasure. After having lived through and survived four recessions in real estate and wine, I felt we were due for another correction. I was tired of always being financially responsible to my employees, to my banks, and to my suppliers. I had created a wonderful unique enterprise, with a collection of some of the best vineyards in the world, but I was ready for a new challenge. I have been told that this is a very typical feeling for a passionate Gemini. When we are done, we know it and do not look back. I did not know what the next challenge might be, but I knew it was time to explore selling.

For the past twelve to eighteen months, I had read in-trade publications that many wineries my size were selling to large American groups, such as Gallo, Kendall-Jackson, Constellation, and Foley. The focus of the wine portfolios were differentiated from low-end to super premium and luxury brands, but what they all had in common were superior distribution systems. They had the pipelines to deliver the juice and the muscle to collect payments—the two biggest challenges of every small producer. Marketing, sales, and collections, occasionally in sync, were often a mix of band aids, chewing gum, and duct tape for artisan winemakers in Burgundy to make and sell enough wine to get to the next year.

With the cynical but realistic admission of my class status (remembering "if you make wine for the classes, you drink with the masses"), I decided to make calls to friends who had sold their businesses. Adam Lee's Siduri, a pinot specialist in Sonoma, sold in 2015 and Lynn Penner Ash in Oregon's Willamette Valley sold in 2016—both to Kendall-Jackson. Adam and I met in Atlanta years before at a wine festival. Although we made stylistically different wines, his richer and bigger, primarily due to Sonoma's fruit profile, his passion nearly surpassed mine as he extolled certain vineyards all the while throwing in technical jargon that even lost me and finishing with "Do you like it?" Lynn also exuded a passion and an intensity for her wines that accelerated as she did tastings. But she was much more circumspect than Adam or me. If I was the straight man for Jim Clendenen, I was her wild counterpart at tastings. I always said that if I was going to make wine in Oregon, it would be modeled on Lynn's wines. If Lynn was on the right and Adam on the left, I would be in the middle—a good place to mine information from both sides. I didn't expect earth-shattering revelations, but I wanted to know what their motivations were when they sold? What did they like and dislike about the business? Where were they personally in their lives and in their careers? What, if anything, did they want to do next? Their responses, in large part matched mine:

- They didn't have adult children interested in working in the business.
- They were frustrated with sales and distribution: more and more time and money spent on sales, marketing, and collecting receivables.
- They were reticent to expand and try new projects (vineyards) due to a dearth of capital.
- They spent less and less time on what they were most passionate about and the reason we all entered the profession: to make wine.

Bottom line, the business was getting less and less fun as we all got older. I remember using that word early on in my career and my dad looking at me askance: "work is not supposed to be fun!" What I meant was satisfying. Work can be pure drudgery and, of course, there are always parts of any business that are less pleasant than others, but when the scales tip to frustrations and satisfaction loses, the gravity of the situation requires pause.

My COO, Alexandre, had made it clear that he wanted to buy into or buy all of the business and I was thrilled. I was even willing to sell the business at a slight discount to him, assuming it was a market price so that I did not have to employ bankers, high-priced lawyers, and the like. A direct sale that eliminated the chattering masses and subtracted the hassle of an auction had a great deal of allure. So I told him that I would not open the sale until we had fully examined his offer and exhausted the possibility of my selling to him.

In February 2018, we met in New York with Alexandre coming from France and I from Jackson, Wyoming, to begin a protocol and discuss his ideas of value. Sadly, our very first discussion made it clear that he would never buy the company. His initial numbers were at book value and I told him I would not sell the business at book—never. Our meeting ended and I encouraged Alexandre to get a firm offer to me and assured him I would exhaust all avenues with him, but repeated that he needed to be competitive. I wanted to get this done before the end of 2018.

Before leaving New York, I met with one of my oldest friends and first partner, Brett Harwood, and discussed the sale. He successfully sold his family parking business twenty years before through an open competition. He braced me for the process, but encouraged me to follow my instincts and begin by looking at taxes. This duality of being a tax resident in France and in the States did not benefit me much as I had to file for twelve-month Visas and create US tax returns for each of my French companies, pleasing my US accountant but not making the US Treasury one cent richer. Examining French and US tax implications in the States was the first concrete step I took. I then considered the values of my various companies and how to enhance their sex appeal (sex and taxes, rather than death and taxes).

Alexandre and I continued discussions but getting a firm offer from him, that had financing, was not forthcoming. I realized that I had the makings of a real problem on my hands with Alexandre running the business, me wanting to sell to him, and him adopting my favorite rope-a-dope move in order to get me to cave in. In early March, I decided to pull my own rope-a-dope and told him: "We will be back in Burgundy in a month. We can address the sale then. In the meantime, let's both focus on the business and wine sales."

Diana and I were heading to a fabulous wine event in Carmel, California, in a week and this would allow me to take the temperature of the market, meet our new importer, and sort out a few thoughts while in one of the most beautiful settings in the world. I won't torture you with the details, but the event felt more like a reunion with old friends from France and the States than work. Jim Clendenen and I did a seminar again, and I reconnected with restauranteurs who I had sold wine to when we both were starting our businesses. Unfortunately, Diana was not feeling in form. But she powered through the events with not only resilience, but enough relish for the two of us. The last meal for twenty high rollers was the definition of over the top: a 10-course marathon that sent me back to my November resolution.

We returned to Jackson to cleanse and think about our retour (return to France). For months, Diana had ailments that, for her, a woman who was never ill, seemed odd. She had passed into menopause but her periods began again and her gynecologist put her on hormonal therapy to regulate this anomaly. Cramps and other GI woes were thrown into the "it's just menopause" camp. However, the "it's-just-menopause" diagnosis did not seem correct, so she scheduled a colonoscopy hoping that it would reveal something that was easily explainable and fixable.

Her doctor arrived after the procedure with the look that does not come close to the actual horror, fear, and dread of his saying: "You have a major blockage in your colon that is cancer. We need to immediately do a CT scan to see if it has spread." I grabbed Diana with a feeling of smallness that was incomprehensible. I was literally compressed in size as the weight of the diagnosis suffocated us, reality became slow motion, and our emotions accelerated to panic speed.

Those first hours in the hospital remain a blur as Diana was wheeled out for the CT and the doctor came back with a dire pronouncement: Stage 4 colon cancer had metastasized to both lobes of her liver.

Chapter 29

Une Mouture (A Grind)

I never wished these pages to be an exposé. But I would not have been inspired to write them without the love and support of Diana.

Diana was a World Cup freestyle skier in the late 1970s and early 1980s when the sport was in its infancy and was also known as hot dog skiing. After knee injuries, she retired from competition and ran the Killington Mountain School's freestyle program from 1988 to 1998, where she developed Olympians. She finished her ski career as a coach of the Salt Lake City Olympic team in 2002. Diana spoke beautiful French and in her off season, she would lead bicycle trips to the wine regions of France. This is how we met in September 2002. After my divorce from Nancy, we were married in 2007. And in the depths of the recession, she launched her own bespoke travel business, The Hidden France, that became a terrific success.

There have been more than enough books written about cancer and I am not about to begin to relate Diana's battles, except to say she organized her treatments as impeccably as she organized her trips to France and her ski teams. She entered this battle with a relentlessness that the oncologists at Huntsman Cancer Institute at the University of Utah and Centre George François Leclerc in Dijon said they had never experienced. Diana's chemotherapy treatments began within five days of her diagnosis. Needless to say, I put on hold all sale discussions. As our lives began to center around her bi-weekly chemotherapy sessions, I had hoped that we would be able to take bucket-list road trips, but the effects of the chemo were fast and nasty. The chemotherapy reduced in size the colon tumors through cancer trench warfare, not the smart bombs of immunotherapy.

I restarted general discussions of selling in May and included several of my closest investors to develop a sales protocol that I would flesh out in the coming months. In June, I flew back to France to lead a trip for Diana who was incapable of traveling, see my team (overdue), get caught up on operations, and talk to Alexandre. As I prepared the sale in my mind, I did not realize the complexity of gathering and assessing the information. For such a small company in revenues, it was a maddeningly complex collection of overlapping independent companies that were all connected from a tax standpoint. One positive was that I had excellent financials dating back to the formation of the company that did not require updating. The French are very good at accounting: they love to keep track of everything. But six-month financials had to be prepared and the summer holidays (France's four weeks of bliss) had me worried that it may not get done.

The companies to be sold included:

- Two separate vineyard companies (partnerships) with a handful of overlapping partners but each operating independently of one another, primarily comprising the Batard-Montrachet vineyard in one and the Puligny-Montrachet and Chassagne Montrachet 1er Cru vineyards in the other.
- Domaine Alex Gambal, my farming company (farm co.) that leased the aforementioned vineyards.
- Maison Alex Gambal, wine-making company (wine co.) that bought the grapes from farm co., made and sold the wine.
- My winery building, that was leased to Maison Alex Gambal.
- Third-party leases comprising ten separate parcels that had to be inventoried and lease payments and terms certified.
- Another twenty separate parcels in transferable and sellable leases with my partner Christophe Buisson.

It did not yet include employee contracts, third-party suppliers, and who knew what else. Meanwhile, I had a critically ill spouse and a big question: would Alexandre make a serious offer or would he be the proverbial fly in the ointment?

When Alexandre and I met briefly in Beaune, we were too busy catching up on current operations and meeting with lawyers to talk. We agreed to meet in July in New York so that he could take the time to prepare and to present me a concrete proposal. In all candor, my engaging our accountant and notaire was a conscious decision on my part to put pressure on Alexandre to make a serious offer. If we could not come to an agreement, I had another sales strategy and I wanted him to know it.

Over the course of two days in New York, Alexandre argued that the businesses were not worth much. "Vous payez trop pour les baux de Buisson. Les valeurs des négociantes pas meux que sont le valeur comptable, il n'y aucune valeur de la marque et vous valorisez les vignes et les baux au moins à 50 pour cent de trop. Au mieux les valeurs sont le vaueur comptable." (You are paying too much for the Buisson leases. Négociants are only trading for book value at best. There is no value to the brand and you are valuing the vines and leases at least 50 percent too high.) I tried my best to act like a patient father to my detriment.

"Ecoutez, je veux bien fair en sorte que cela marche pour vous, et je vais vous vendre l'enterprise sur plusieurs années. C'est clair que vous ne pouvez pas avoir le financement nécessaire et c'est un moyen pour que nous puissions tous les deux accomplir notre but." (Listen, I am willing to make this work for you and I will sell you the business over a number of years as an installment sale. It's clear you cannot get the financing you need and this is a way for both of us to achieve our goals.)

"Je ne peux pas avoir de partenaires; Je dois tout posséder maintenant." (I cannot have partners; I need to own it all.)

I thought to myself, "Wow, that is heavy. This kid has hutzpah and no sense of reality." I am willing to make this work for him, but he keeps saying I want you to sell me the business cheap and get nothing for twenty-five years of work. "Alexandre, je ne vendrai jamais à la valeur comptable. En fait, je ne vendrais pas à mes propres enfants au livre. Si c'est tout ce que vous proposez, nous devrions cesser les discussions maintenant, car je sais que je peux obtenir plus en commercialisant les entreprises dans le monde entier, et je le ferai." (I will never sell at book value. I would not even sell to my own children at book. If that is all you are going to offer, we should cease discussions now because I know I can get more by marketing the businesses worldwide, and I will.)

"Vous ne le vendrez jamais pour ce que vous pensez qu'il vaut." (You will never sell it for what you think it is worth.)

"C'est un risque que je suis prêt à prendre. Je ne vendrai à personne au moment du livre, je suis un acheteur à la valeur comptable, pas un vendeur." (That is a risk I am willing to take. I will not sell to anyone at book. I am a buyer at book value, not a seller.) "Alexandre, don't you see that I am offering to help you acquire a terrific business platform that with energy that you can double or triple in size over the next ten years? You are buying some of the past but mostly the future: this is what I have reflected in the value."

"Je ne pourrai jamais doubler ou tripler l'enterprise." (There is no way I can double or triple the business.)

"Alors, vous ne devriez pas acheter l'entreprise. C'est la raison pour laquelle vous l'achetez en tant qu'entrepreneur et non en tant que gardien." (Then you should not be buying the business. This is the reason to buy it, as an entrepreneur not as a custodian.)

At that moment, I knew that Alexandre would never be a buyer and simply said, "Alexandre, even if I did not have a fiduciary responsibility to sell to the highest bidder, I would do so. I have twenty-five years of physical and personal capital invested and I intend to maximize that value. I do not care what you believe the values are. I could be wrong, I have been wrong many times in my life. And if I am, I can live with my mistake. You cannot make money unless you take risks. I will sell to the highest bidder, period. If it is not the number I imagine, what is the worst that can happen?" Alexandre paused and thought but did not respond. After a long pause, I said "Je serai toujours propriétaire de l'entreprise et je continuerai à travailler." (I will still own the business and will continue working.)

I left Alexandre knowing that France would be in suspended animation for the next four weeks and this would give me time to catch my breath, except that I started to make wine in California in 2016. The proverb "idle hands are the devil's workshop" comes to mind. With my team running the daily operations, I was always haunted by Julia Child saying "What am I going to doooo?" in the film *Julie & Julia*. To make a long story short, in 2015, my partner from Santa Barbara and I were discussing sales problems in California. We talked about a new importer, distributor, direct sales, employing agents—any and all were on the table. As I explored the Santa Rita Hills, a wonderful region for pinot noir and chardonnay, I became fascinated with its terroir and was drawn in again by the rocks. Before I knew it, we were digging holes and putting a mini winemaking plan together. With Peter Work and his Ampelos winery, we decided to make a bit of chardonnay in 2016 in the Burgundian Gambal fashion (whatever that means). In all seriousness, I found the vineyards and their management first rate. In fact, we in Burgundy could learn a few things from them. On the other hand, there was a clear disconnect between the vineyard and the winemaking: a color-by-numbers and one-size-fits-all approach was de rigueur. They were taught to pick the grapes when they reached a certain acidity and sugar level to get a specific flavor profile rather than saying, "Wow, the grapes sure taste great today. Let's see what they will give us." I was taught to tune into the grapes Burgundian style: listening to them and to the vineyard and finding opportunity to make great music: admittedly on the fly and risky as hell, but the rewards are why I and others do it.

In 2016, we produced two very successful vineyard-designated chardonnays that clients and the press were very complimentary of, saying it was an atypical chardonnay from California. A Burgundian-style California wine that was "Very Gambal." That made me smile and was my primary goal. The 2017 vintage was more complicated due to torridly hot harvest conditions (that old learning curve) and we made a blend of the same two vineyards with a bit of advice from my French consiglieres, a bit like me pulling a rabbit out of the hat. Perhaps not a slam dunk of a wine in 2017, but a very pretty bank shot nonetheless.

Fast forward to 2018 and we decided to ramp up production a bit and make two pinot noirs along with the chardonnays. In early September, I am selling the business, my wife is terribly ill, and on the eve of harvest my winemaking partner, Peter, becomes bedridden with pancreatitis until early October. It was one of those moments, as a 15-year-old says after getting caught driving and then wrecking his dad's car, "At the time, it seemed like a good idea."

Business is a bit like a battle: the minute you begin the engagement, all the planning goes out the window. In this case, we had a great plan with me, Peter and his team providing guidance. But with his absence, I was flying solo. Luckily, Peter's neighboring winemaker helped me get organized. I don't know how I did it, but I took a break in the middle of harvest to return to Jackson for Diana's thirteenth chemo and third CT scan and I vinified the same two vineyard chardonnays plus two delicious single vineyard pinots noirs in beautiful oak fermenters that I bought and shipped over from my Burgundy.

With my Californian wines safely in barrel, Diana and I could return to France for the end of Burgundy's harvest and for Diana to continue her treatment at the University Hospital Center in Dijon. If there was a harvest of mostly positives and few worries, it was 2018: a year of quality and quantity. I spent October hovering over my flock as they gave me looks of "Patron, tout va bien. Ne t'inquiètes pas." (Boss, everything is going great. Don't worry.) As suspected, the first half of our financials were delivered in October. I continued work with my Notaire Martine, hired outside counsel and Rothschild as my investment banker to market the business.

Alexandre, to his credit, managed harvest well but there was no movement on his part to make his previous offers serious. He had made it clear to me in an earlier conversation: "Ne penses pas que je vais vous aider à vendre l'entreprise au plus offrant." (Do not think that I am going to help you sell the business to the highest bidder.) It was clear that Alexandre could make my life miserable during the sale, but he could not stop it. I could not under French law get rid of him (very expensive and painful), and was totally prepared to stay in France until the sale was completed and to continue Diana's treatments in Dijon. Diana and I were ready to do what was necessary to affect the sale with a bit of liar's poker thrown in for good measure.

In late October, Alexandre and I met in Paris and I encouraged him to bid (or not)—even though his offer was not serious. I was also certain he would stay, a name-your-poison choice for me because:

- He needed the money (I knew he was going to start his own winemaking business)
- He wanted to make my life difficult, but he could only go so far even under French law being a nudge
- He figured he would have some leverage on me to get a better exit package

- He really never thought I would call his bluff and come back to Beaune and take over

Our meetings were short and at the second one I asked point blank "As-tu changé d'avis concernant l'offre et ton rôle pendant la vente?" (Have you changed your mind on making a different offer and what your role will be during the sale?)

"Non" he replied, as he got up to walk out.

"OK, je te verrai demain après-midi au bureau après mes réunions à Paris." (OK, I will see you at the office tomorrow afternoon after my meetings in Paris.) I could see the shock on his face that I called his bluff. He never thought that I would return to Beaune and would be prepared to set up shop until the sale was completed.

Chapter 30

La Ligne d'Arrivée
(The Finish Line)

I closed the sale of the business on July 24, 2019, during another one of France's now more frequent heat waves. The actual closing was without incident if you call twenty-five people in a conference room with no air conditioning on the outskirts of Dijon where it was 100 degrees outside uneventful. We were in a chic, new green office with no freaking air con: unbelievable. To add insult to injury, the building had no high-speed Wi-Fi. In France, all real-estate documents are now signed and registered electronically but no Wi-Fi signal, no deal. I nearly blew my top (péter un cable). At the head of the table, I observed this circus on one level as a bemused third-party observer, but the sweat dripping down my back reminded me that this was my freaking nightmare. The notaire tried vainly to create a hotspot on his phone that was propped next to the window for better reception. Three hours into the signing, which was proceeding at a Burgundy's snail-mail pace, I started to tap the plastic electronic pen harder and harder on the table and kept telling the notaire, Martine's son Felix, "Signons ces putains de papiers et tu pourras les engregister plus tard!" (Let's sign the damn papers and you can register them later!)

"Attend, ce n´est pas ma faute, nous sommes bientôt là." (Hold on, it's not my fault, we are almost there.) In France, it is never my fault. Perhaps I made an error (one erreur de ma part), but it is never my fault. Somehow toward 7:30 with an 8:00 pm dinner reservation looming, a connection magically appeared, we signed, and now could celebrate with a toast!

As the seller, I was given the honor (and obligation as I found out), to give the closing toast. I said: "Je tiens à remercier toutes les personnes présentes dans la salle pour leur patience et leurs efforts à parvenir à clôturer cette transaction. C'était une structure extrêmement compliquée mais il est clair que tout le monde dans la salle a vu ses mérites et donc je ne pourrais pas être plus heureux que la famille Boisset ai acquis un merveilleux outil pour continuer à faire des vins encore meilleurs que ce que j'ai pu faire pendant ces vingt-trois ans merveileusses." (I want to thank everyone in the room for their patience and focus on bringing this transaction to a close. It was an extremely complicated structure, but it is clear that everyone in the room saw its merits and thus I could not be happier that the Boisset family has acquired a wonderful new instrument to make even better wines than I have during these wonderful twenty-three years.)

As we drove to dinner, I had a full range of emotions. First, incredibly sad that Diana would not be with me at dinner. She was ill from a trial-run radiation treatment on her liver with the full treatment scheduled a week later. It was Diana who was with me from the dark days of 2003 to 2005 with many Boston and Namur moments. She deserved the credit for my success.

I thought of Alexandre, who did stay until the final sale and I had accepted his attitude as the lesser of two evil choices: fire him and deal with France's notorious labor courts that virtually always side with the employee, or live with him, ignore this open sore, and continue the rope-a-dope, knowing that he was going to leave with a generous severance package. Years later, I still feel his exit package was too much and I am sure he felt it was not enough: a sign that our agreement was probably fair for both parties.

My lead attorney, Laurent Bensaid, was an amiable Frenchman in his mid-40s with a soft voice and a charm that aggrandized his size and presence. He became a dear friend whom I affectionately refer to as "l'écraseur" (the crusher). I have never met someone who could take a position on my behalf, one that I thought was ridiculous and then after five minutes have me convinced he was correct. He would say plaintively to the opposing attorney, "Mais noooo" (But nooooo), and I would fasten my seatbelt because I knew we were in for a rhetorical flourish that would have made Alcibiades, the cunning Athenian who switched sides in the Peloponnesian War, proud.

In early June, there was a surreal moment that still stands above all others. As the deal's final details were being hammered out, I could not attend the negotiating session because Diana was receiving chemo number 44 in Dijon. It was good that I could not attend because in most negotiations where performance language (word parsing) is being hammered out by the attorneys (one thing that's the same in France as in the US), preventing an entrepreneur from saying something stupid out of frustration to the lawyers is a good idea. It was agreed that the principals would get on a conference call that morning to review the lawyer's work and ideally come to an accord over the disputed points, the principal being a breakup fee. In my heart, I knew the buyers wanted to close on the deal as much as I did. I was certain of this. But of course, our respective lawyers assumed this was not the case and this is why they get the big bucks. As the conference call continued, I was talking on the phone with the various participants and simultaneously texting Laurent to make sure we were in agreement, all the while in my car in the parking lot of the cancer hospital in Dijon. I thought to myself, "This in unbelievable. My wife is fighting for her life and getting sick from her umpteenth chemo infusion and I am in our car in the freaking hospital parking lot trying to sell my life's work. Is this real?" As the figure for the breakup fee was agreed to (I actually texted Laurent to tell him to make a deal at a number while he wanted to grind for more), I thought, "I hope the deal does crater and they walk. With the cash they would owe me by bailing, I can pay off all my debt and have two to three years of found cash flow!"

Our closing dinner was bittersweet for sure without Diana. But the wonderful American-strength air conditioning was a delight.

Tasting notes can be both numbing and simply boring as hell. But I will always remember the wine notes from our dinner. The wines we drank (not tasted) reeled me back to another dinner thirty-five years before, when I first really began to taste, drink, and learn about wine. I will admit that I was spoiled, knew it, but it only increased my curiosity. Through that curiosity, I learned that wine is nothing more than fermented grape juice, so what is the big deal? Why get all excited over a fermented beverage that wants to make itself? What makes people swoon over a bottle of Romanée de la Conti but turn their noses up at a bottle from a vineyard 400 meters further east? Why are certain bottles and vintages legendary and others not? As with most things, the answer lies in the eye of the beholder. Are you open to the transformational magic?

I am not a golfer, but I have many friends who can tell me the distances of holes at Augusta or Pebble Beach, which one is more difficult and how best to play the hole. I have no clue what they are talking about and I would never consider paying hundreds of dollars to walk around and hit a white ball. I can exercise by walking out the front door wherever I am. I love living in the mountains and sharing my love of wine with others after a great day of skiing. To be sure, these activities do make our lives more interesting and richer. But let's never forget we can live without them.

I have had some transformative bottles and they were because of the generosity of my early wine circle. The greatest was, and always will be the 1947 Musigny Vieilles Vignes (old vines) from Comte Georges DeVogue that I drank in magnum in the winter of 1993. The late Becky Wasserman and several of her colleagues were visiting Washington, DC, to sell wine and to meet with me to plan my exodus from Washington a few months later. Larry Smith, who has also passed, was our host. That night, he wore baggy sweatpants that revealed his plumber's crack as he stoked the fireplace. After sharing with us several old and fantastic bottles, we finished with Le Musigny VV 1947 in magnum. This was and still is quite simply the greatest wine I have ever touched to my lips and I was oblivious at the time about its legendary stature.

The perfume, no, literally the wine itself wafted out of the Riedel goldfish bowl-sized glass. In a sense, the wine and its perfume were the same thing. The wine was black, dense, rich, not heavy but with the taste of fresh-picked and perfectly ripe grapes. I still recall the taste as though it was yesterday.

The closing dinner wines were delightful. A thirst-quenching champagne from a small producer and some of mine and Boisset's signature bottles. My white was Puligny Les Enseignères 2014, the best vineyard and bottle I made for the money and my red was my 1999 Clos Vougeot, still far too young. Their white was Vougoet 1er Cru Clos Blanc, a lovely rich, elegant chardonnay planted in white soil surrounded by some of the greatest red-soil pinot noir vineyards in the world. Finally, the Boissets served a magnum of 1999 Musigny from their Domaine de la Vougerie that took me back to that bottle of 1947 Musigny I had at at the beginning of my career. The 1999's DNA was the same as the 1947: opulent with a perfume that exploded with a large spray from the glass, an elegance, power, and balance like no other wine in the world.

I wondered, how could two wines made fifty-two years apart be so alike? I was no closer to the answer than I had been in that room full of blueberries at the Four Seasons Hotel twenty-six years before in Washington, DC. But I am content to know that the question is still the same and the answer yet to be found.

Afterword

I wrote a great part of the preceding pages in the spring of 2017 just after our second consecutive year of frost (the one averted by burning hay bales). I remember it well because I coughed as I wrote. I did not expect that my instincts about an economic storm would be realized by the events of the next three years.

In the fall of 2019, the Trump administration put in place a 25 percent tariff on alcohol in retaliation for European Airbus subsidies. Aimed primarily at the politically sensitive wine industry of France, Germany, Spain, and England, this action was a shock to the system and greatly reduced exports to the States. That fall, importers, distributors, and especially growers had to lower margins in order to maintain a semblance of sales. The tariff was not repealed until May 2021.

In January 2020, when I returned to France to help Diana with her trip, I attended the funeral of Michel Lafarge. He was one of my mentors and I consider his son, daughter-in-law, and their children my family.

In March, Diana's illness became my full-time job as it became clear that the cancer had taken over, despite her refusal to give in to it. She maintained a positive, upbeat attitude even after fifty-five chemo treatments, when other tumors appeared and began to overwhelm her.

That winter, COVID-19 stopped all trade in its tracks. A partner, supplier, and organizer of trade shows also sold his business in the fall of 2019, and we celebrated together as commerce grinded to a halt. Diana and his wife Kim teased us. "We do not know if you guys are that smart or just plain damn lucky." We both agreed while laughing that we were not that smart and sometimes you just hit the number.

Lockdowns were the norm, but they were the least of my worries. In Wyoming, I was able to get outside and ski in the back country with friends to keep myself grounded. I learned a great deal about caregiving, the saints that are hospice nurses, and the strength we all can have.

Diana died on June 18, 2020, after a twenty-seven-month fight with cancer. Despite this, or perhaps because of it, her spirit became stronger. Even when she was wracked with pain, she was determined to live. Her death I accepted. Her suffering broke my heart.

It is hard to remember the days after Diana passed. I wandered around in a bit of a haze as the finality of her absence set in. My daughter, Alexa, her husband, Grant, and their Golden Retriever Maggie eased my suffering as they moved in while their home was being finished.

That next winter, I forced myself to get out and move on from my funk (grief never ends), aided in no small measure by my friends in the Jackson community and the ability to be outside and work on the Jackson Hole ski mountain as a host or what I affectionately call the JV ski patrol.

In April 2021, France, and especially Burgundy, was hit with a series of freezing mornings that destroyed 60 to 80 percent of the crop. The vignerons were ready and lit smudge pots/candles to battle the cold, but the depth of the cold and its duration were unprecedented. Photos were published, quite beautiful in fact, of the vineyards in candlelight but that belied the damage that was occurring. The reasons for the damage were similar to that of the 2016 frost: a mild winter, very early spring, and a far too advanced growth stage that exposed well-developed buds and infantile grape bunches to the freezing temperatures. My friends in Burgundy said, "The timing of your sale was perfect."

I finally returned to Burgundy in June 2021 as a vaccinated tourist, three days after the borders opened. Having not been home for almost eighteen months, I was thrust into the process of cleaning up and clearing out our home of Diana's belongings. It was the hardest two weeks of my life.

After the sprint of harvest, there is a pace in Burgundy that reminds me of a quartet movement in adagio, a pace that is neither pressed nor slow. It is wonderful to be able to return now with this pace.

I returned again in the fall for a month to try to get my bearings and felt a bit more settled. It was a lovely visit as I traveled over France visiting old friends and other winemakers. I decided to turn my home into a vacation rental when I was not there in order to keep its maintenance in good order.

Now when I come back to my home in Orches, I have the pleasure of tasting wines with the people who worked for me over the years. They have all had great success in their careers. Fabrice Laronze, who started to work with me in 1998, left my employ in June 2009, and created his own critically acclaimed domaine, Domaine de la Velle, with his wife, Sophie. It was during Fabrice's years that we received our first star in the RVF, *La Revue du Vin de France*, the French wine equivalent to the *Michelin Guide*. Geraldine Godot joined Domaine Gambal after Fabrice and worked with me until 2014, when she moved to become the regisseur (head of winemaking) at Domaine de l'Arlot, in Premeaux-Prissey (Nuits-Saint-Georges), where I did my apprenticeship in February 1997. She has taken Domaine De L'Arlot to three-star status in the RVF. Megan McClune, who had a variety of roles over eleven years and always kept my financials straight, moved to Domaine Jessiaume in the spring of 2014 as its managing director. She totally restructured the domaine, hired their winemaker who did his apprenticeship with me, and now is the general manager at Maison Benjamín Laroux.

Matthieu Thevenard, who also did his apprenticeship with me in 2013, replaced Geraldine just as he graduated from wine school in June 2014. He was reluctant to take the position because he did not think he was ready, but I assured him that he was much more technically qualified than me. I watched over his shoulder. As Diana would say, "He will become a big fat success." He has developed with his family's vineyard holdings a domaine in the Mâconnais. His first wines in 2021, a year I cannot imagine could have been more difficult, are terrific. They, along with all of my mentors in these pages, are my Burgundy family and I am delighted to witness them as they continue to raise the standards of our beloved vignobles (vines).

I was overjoyed to attend the marriage of Matthieu and Camille (an accomplished winemaker in her own right), and to help Megan's husband, Matt (yes, Matt of the signs), explore the nuances of another crop by creating a coffee business.

Becky Wasserman, who opened Burgundy's door to me, passed in August 2021. Her export business is thriving with her sons on the scene discovering and promoting up and coming vignerons. Alexandre, my COO, went on to create his own domaine in Mersault and was having great success when he tragically succumbed to pancreatic cancer in September 2022. He was the same age as my son, Tyler "Coyah," and left behind a wonderful wife and two young beautiful daughters.

In the summer of 2023, Coyah and his bride, Zyra, went to Orches for their honeymoon. My daughter, Alexa and son-in-law Grant, began raising my beautiful grandson, Will, a few blocks away from me in Teton Valley, Idaho. I finished these pages in the fall of 2022, after my partner and I spent six weeks in Burgundy editing together. Her skills are formidable and the success of this book is hers as well. I could not have completed it without her. Burgundy and its winemakers are smiling. They had a hot dry 2022 summer but a full and qualitative harvest: their cellars are full. The dollar is at its strongest in a generation and quite frankly everything seems cheap in France, as compared to the States. What I call the grocery cart test, shows that the cost of a similar cart in the US is easily 30 percent more expensive (my gut tells me it is even more and I am not talking about going to Whole Foods). This nonscientific study does not even factor in the quality of the meat and vegetables that are of a much higher quality. My mouth waters as I see the various cheeses, chickens, rabbits, sausages, fresh fish, and Charolais beef, which you can now get dry aged on display in France.

It has been astounding to observe the escalation of Burgundy's prices. It is clear that for a variety of reasons, the inelastic nature of Burgundy's pricing structures is changing and the price for its wines is skyrocketing, at least for the moment. A strong dollar and low supplies both at the producer and in the supply chain are part of the answer but they cannot tell the whole story. Quite simply Burgundy is the "it girl." At the moment, it is hot and a great many people with money in their pockets are willing to pay for it.

The handful of collectibles continue to climb to what I consider stupid levels. I remind my friends, "It is just grape juice" but more power to you if you have the resources to buy them. And more power to my other friends who can sell their wine at these levels. At the lower levels, it is frankly surprising to see wines that I bemoaned not being able to sell for a correct profit now selling for 35 percent more than my wholesale pricing just three years ago. This race to and for the top has allowed the more modest wines to draft along in their wake to more profitable levels. Then there is a third group, the new celebrity winemakers that everyone must have and, in turn, are pricing their wines at the emperor-has-no-clothes levels. A London merchant, one of the oldest and a supplier to the royal family, told a young winemaker in Vosne-Romanée (in polite English, I'm sure) that he could keep his wines and find, as P. T. Barnum would say, another sucker to buy his wines.

Is this a bubble or a new new? As long as our end customers have money, or think they have money in their pockets, I suspect it is a new base from which all the elements of Burgundy will benefit. However, if there is a string of large harvests and a severe economic contraction another outcome is likely. Only time will tell.

One trend does trouble me as it relates to the big money that has suddenly found Burgundy attractive. Burgundy's vines have never been cheap but there has always been a mix of locals and some outside wealth looking for a generational investment in or with a local domaine. Not to throw stones because I too was probably accused of overpaying, but it seems that unimaginable wealth has come to the landscape. In 2022 alone, a domaine in Vougeot, with spectacular holdings was bought by an American group from Oregon. Bouchard Pere & Fils was sold in 1995 to Henriot, a not-insignificant Champagne house, then again to Francois Pinault, one of France's richest men. Two estates in Pommard have been bought by Americans who, to their credit, have and are investing extraordinary sums in building renovations and vineyard and winemaking improvements to properties that had languished for years.

The Hospices auction of November 2022 was wildly successful and it raises the specter of the aforementioned bubble. Not only did prices break all previous records, but the total raised, nearly 29 million euros, an extraordinary sum for any institution much less a local hospital, makes me pause. Yes, the auction is for charity, but when we in the trade consider prices for wine from good, yet modest, vineyards selling in barrel for the equivalent of 55 euros per bottle before any élevage, bottling, labeling, taxes, and shipping (adding an additional 50 percent to the cost), Burgundy has entered another world. Yes, the wines are rare and, yes, they are delicious, but collectible art they are not. The Grand Crus sold at over 500 euros per bottle, with some at 1,000 euros, but the most mind-bending statistic of all is that the average price of the nearly 230,000 unfinished bottles sold was 125 euros before costs.

This said, I was lucky to buy two barrels to benefit the Dee Williams Freestyle Fund, a foundation that I set up in Diana's memory that gives scholarships to up-and-coming freestyle mogul skiers.

The feeding frenzy, and I use these words purposefully, has reached levels that to me are purely speculative. Tulips (1634), The South Sea (1720), Japanese Real Estate (1984), dot com (2000), and the Housing Bubble (2008) come to mind.

I'm just glad for stories that show money can't always make a deal in Burgundy. A good friend in the Côte de Nuits recently secured a spectacular long-term lease on a parcel of Chambolle-Musigny 1er Cru from an old-line merchant family in Dijon, whose deceased patriarch had bought the land many years ago, loved it, and told the family never to sell it. It was a piece of their heritage. Even though they had a who's who of Forbes interested in the parcel, and they could have inflated the price because of the feeding frenzy, they decided to keep it local. They wanted to have a relationship with their winemaker tenant who they felt would do the right thing with the parcel, be part of their extended family and add to the heritage.

I am grateful for my time in Burgundy and the joy Diana and I shared there. Now, when I return, the acute pain of her death has passed and I am on the mend. I walk through the Saturday market with wonder and ease, catching up with old friends. The magic of Burgundy continues.

About the Author

Originally from the Washington, DC, area, **Alex Gambal** moved to France in 1993 and, over the course of the next twenty-six years, straddled the Atlantic while creating a boutique winery that produced wines from some of Burgundy's greatest vineyards. A hard-nosed businessman in a community steeped in tradition, he defeated the odds by acquiring some of the most-coveted French vineyards and creating a successful brand that was eventually sold to one of the largest winemakers in the region.

He spends his winters in the Tetons, where he is a ski host at the Jackson Hole Mountain Resort in Wyoming and an avid hiker. He is also the benefactor and chairman of the Diana "Dee" Williams Freestyle Fund, a charitable foundation to benefit the development and advancement of promising mogul skiers to the Olympic and World Cup level.